Pilot Protective Relaying

ELECTRICAL AND COMPUTER ENGINEERING

A Series of Reference Books and Textbooks

FOUNDING EDITORS

Additional Volumes in Preparation

Pilot Protective Relaying

edited by

Walter A. Elmore

Blue Ridge, Virginia

(Retired)
ABB Automation, Inc.
Substation Automation & Protection Division
Allentown, Pennsylvania and
Coral Springs, Florida

CRC Press
Taylor & Francis Group
Boca Raton London New York

CRC Press is an imprint of the
Taylor & Francis Group, an **informa** business

ISBN: 0-8247-8195-3

Headquarters
Marcel Dekker, Inc.
270 Madison Avenue, New York, NY 10016
tel: 212-696-9000; fax: 212-685-4540

Eastern Hemisphere Distribution
Marcel Dekker AG
Hutgasse 4, Postfach 812, CH-4001 Basel, Switzerland
tel: 41-61-261-8482; fax: 41-61-261-8896

World Wide Web
http://www.dekker.com

The publisher offers discounts on this book when ordered in bulk quantities. For more information, write to Special Sales/Professional Marketing at the headquarters address above.

Preface

The current volume, *Pilot Protective Relaying*, is complementary to *Protective Relaying Theory and Applications* published in 1994 (Marcel Dekker, Inc.). *Pilot Protective Relaying* covers the basic fundamentals of high-speed transmission line relaying. As in the first volume, the fundamental aspects of protective relaying are emphasized without concentrating on the means of implementation. Simplified consideration is given to the communications channels that are a vital part of these types of systems. Chapters are included describing three-terminal applications, program design for microprocessor relays, protection of lines equipped with series capacitors, and single-pole relaying. Information is included on substation automation, the nature of IEDs (Intelligent Electronic Devices), and what is involved in information management for a substation. A chapter is included on digital fault recording, with emphasis on its use as a relay design tool and as an aid in relay performance analysis. The bibliography lists many recent papers that will prove useful to relay engineers.

At this point in the history of protective relaying, the fundamentals, which have long been handed down from generation to generation, are in danger of being lost. In the past, each utility, industrial company, or manufacturer had a cadre of relaying experts, with one leading central figure who had devoted his life and the majority of his waking hours to the investigation and study of relaying nuances and problems. The development and continuing presence of such dominant figures, with their guidance and training of their fellow engineers and technicians, are in danger of being sacrificed in the interests of flexibility and the ever-important budgetary bottom line.

It is with this thought in mind that the authors of *Pilot Protective Relaying* have attempted to lean on the side of simplicity and root technology, rather than stressing the most recent esoteric refinements of the established art. It is hoped that this book, along with its companion volume, will provide a guide to the basic principles involved. Other excellent sources provide access to much of this information, but in general they assume previous knowledge that may not exist, provide at best a transient display of information, and are often written in unintelligible language that few can interpret.

All the authors of this book have devoted many years to their specialties, and their knowledge has been honed in the fire of everyday practical exposure to the perversity of nature. Each has contributed measurably to the continual improvement of protective relaying and substation control.

Many of the long-established principles, which are described casually in this and in its companion volume, owe their existence to the brilliant and talented engineers who originally created or refined them. A partial list of those who made a distinct impression on these two volumes and, in some cases, have received inadequate recognition for their influence on the broad field of protective relaying are: C. L. Fortescue, C. F. Wagner, E. L. Harder, R. C. Cheek, W. K. Sonnemann, J. L. Blackburn, W. E. Glassburn, G. D. Rockefeller, Finn Andersson, H. W. Lensner, Roger E. Ray, W. L. Hinman, Hung Jen Li, and E. A. Udren.

Walter A. Elmore

Contents

10. Protective Relay Digital Fault Recording and Analysis   137
Elmo Price

Biographical Sketches

Walter A. Elmore was born in Bartlett, Tennessee, served as a navigator in the Air Force, and graduated from the University of Tennessee with a B.S.E.E. in 1949. He was in Substation Design at Memphis Light Gas & Water Division until he joined Westinghouse Electric Corp. in 1951 as a district engineer in Seattle, Washington. He transferred to the Relay-Instrument Division in Newark, New Jersey in 1964 where he became manager of the Consulting Engineering Section. He held that position, following a 1989 merger with ABB, until 1992 in Coral Springs, Florida. From 1992 until his retirement in 1996, he held the position of consulting engineer. He continues to consult for ABB.

Mr. Elmore is past chairman of the IEEE/PES Technical Council, past chairman of the IEEE/PES Power System Relaying Committee, and a Life Fellow of the IEEE. He is a member of the Phi Kappa Phi, Eta Kappa Nu, and Tau Beta Pi honorary fraternities. He received the 1989 IEEE Gold Medal for Engineering Excellence and the 1989 Power System Relaying Committee Award for Distinguished Service. In 1996, the ABB manufacturing plant in Coral Springs, Florida, was dedicated to him. In 1997 Texas A & M presented him with the Most Prolific Author Award. He was accepted as a member of the National Academy of Engineering in 1998. He has presented over 100 technical papers, holds six patents, and is the editor of the book *Protective Relaying Theory and Applications*. He is a registered Professional Engineer in the state of Florida.

William J. Ackerman started work with Automatic Electric Company on the first solid-state SCADA systems (CONITEL-2000) after receiving his B.S. and M.S. degrees in electrical engineering. He later worked for Leeds & Northrup Company and Florida Power Corporation before joining ABB as a Project Manager in the Systems Control Division. He transferred to the Power Automation and Protection Division in 1996, where he is currently Manager of Substation Automation Systems. He is a Senior Member of the Institute of Electrical and Electronic Engineers (IEEE) and the Power Engineering Society (PES). He is past-Chairman of the Substations Committee of the PES, and of the Automatic and Supervisory Systems Subcommittee. He is a member of the IEEE and International Electrotechnical Commission (IEC) committees and working groups concerned with the development of substation automation standards. He has authored and coauthored numerous papers, including the IEEE Tutorial "Fundamentals of Supervisory Systems."

Elmo Price received his B.S.E.E. degree from Lamar University in 1970 and the M.S.E.E. degree in power systems from the University of Pittsburgh in 1978. He began his career with Westinghouse in 1970 and held several engineering positions, including in design engineering and marketing for the Small Power Transformer Division, electrical systems design engineer for the Gas Turbine Systems Division, T&D product application support for T&D Systems Engineering, and product application engineer and advanced technology specialist for T&D Marketing. With the consolidation into ABB in 1988, he assumed regional responsibility for product application for the Protective Relay Division. In 1992 he joined the Substation Control and Communications group in Coral Springs. In 1994 he became the Manager of Product Management and Consulting, which is responsible for protective relay application support and defining new concepts for system protection to meet new market and technical requirements. Mr. Price is a registered Professional Engineer and a member of the IEEE.

Liancheng Wang obtained his B.S. and M.S. degrees in Electrical Engineering from Shandong University of Technology, Jinan, China, in 1983 and 1986, respectively. He obtained his Ph.D. degree in electrical engineering from Clemson University in 1995. From 1986 to 1991, he was an assistant professor in the Electrical Engineering Department, Shandong University of Technology. He joined ABB's Power Automation and Protection Division as a senior engineer in 1996 and currently serves as Technical Director for High Voltage Applications. His areas of interest include microprocessor relays, signal processing, and power system stability. He is a Senior Member of the IEEE.

1

Communication Fundamentals

WALTER A. ELMORE

1. INTRODUCTION

This volume is complementary to *Protective Relaying Theory and Applications* and covers pilot relaying extensively. Inherent in pilot relaying is the need for a reliable communications channel. This chapter does not purport to describe the details of modern relaying communications equipment, but rather discusses the basic principles on which much of it is based.

2. BASIC EQUATION

Much communications technology is based on the simple equation:

$$V = A \sin(\omega t + \phi)$$

Three bits of information are incorporated in this expression: the peak magnitude of the sinusoid, the frequency, and a phase angle.

3. ON–OFF SIGNALING

By causing a transmitter to change its output signal magnitude in response to a relay function, a receiver at some remote distance is able to detect this change and to initiate the appropriate action. The extreme variation of A

utilized in power line carrier introduces a change from zero to full magnitude. This is described as *on–off* carrier.

Figure 1-1 shows an example of on–off keying for a power-line carrier channel. The quiescent state is "off." The oscillations are at a chosen frequency in the 30 to 300-kHz range. An extensive description of the application of a power-line carrier is covered in Chapter 3.

4. FREQUENCY SHIFT KEYING

A frequency change (known as frequency shift keying; FSK) is associated with "frequency-shift" channels. With this system, there is a normally transmitted frequency called "guard" or "space." This concept is used with audio tones as well as with a power line carrier. As relaying action dictates, the transmitter frequency is shifted to a different (usually lower) level, identifying to the receiving location the need for relaying action. This signal is called "trip" or "mark." In general, the terms guard and trip are used if the function of the channel is to produce circuit breaker tripping directly. Space and mark are generally used when the functions of the two frequencies are essentially equal, such as in a phase comparison relaying system, where tripping may be accomplished in response to either frequency, provided other conditions are met.

1

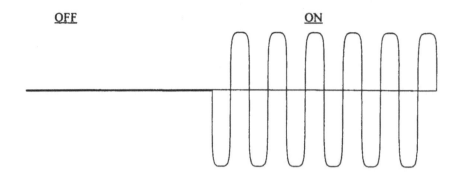

Figure 1-1 On–off signaling.

Some frequency shift systems operate at a lower power output level in the guard state, and shift to a higher power level when in the trip state. Transmitting at a lower power level during normal, prolonged operation allows the detection of a non-trivial reduction in received signal level. Detection of an impending channel failure is then detected while keeping in reserve the higher power level when tripping is desired.

The shift of frequency initiated at the transmitter and detected by the remote receiver allows a logical decision to be made for the need of a relaying system to operate or to refrain from operating. Figure 1-2 shows an example of the waveform associated with a frequency shift channel. The slight change in frequency is detected by the receiver and appropriate action is taken by the relaying system.

5. PULSE–PERIOD MODULATION

An important variation of the frequency shift concept is one in which a continuous variation of frequency is accommodated. This is called a pulse–period modulation (PPM) scheme.

This scheme was devised to allow a 60-Hz (or 50-Hz) sinusoidal signal at one location to be completely reproduced at a remote location in the interests of permitting a direct comparison in magnitude and phase angle of that waveform to a similar sinusoid generated at that remote location. This allows a pure differential concept to be used for transmission line relaying in a manner comparable with that used for generator relaying. The complete scheme is described in Chapter 4.

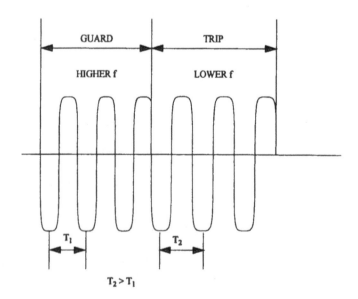

Figure 1-2 Frequency shift signaling.

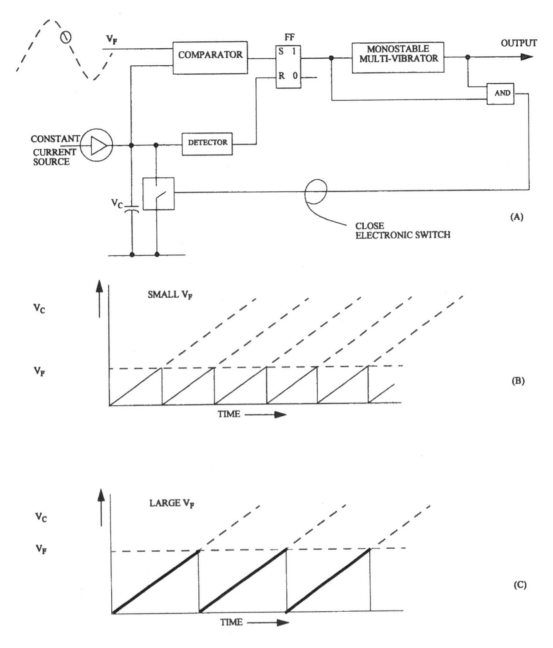

Figure 1-3 Pulse–period modulation.

Figure 1-3A describes the concept used at the transmitting terminal. The single-phase sinusoid is generated as described in Chapter 2. This composite filter output is referred to as V_F in this figure.

The constant current source produces a constantly rising voltage V_C across the capacitor. When V_C reaches the level of the instantaneous value of V_F the flip-flop is triggered, causing an output to occur from the nonstable mul-tivibrator for transmission and also causing a temporary short of the capacitor. The cycle then repeats, producing an output similar to Figure 1-3B or C. The higher the magnitude of V_F the longer the time required to change the capacitor to that level. The output frequency, then, is inversely proportional to the instantaneous value of V_F (more precisely the period of the output is modified by a constant times V_F).

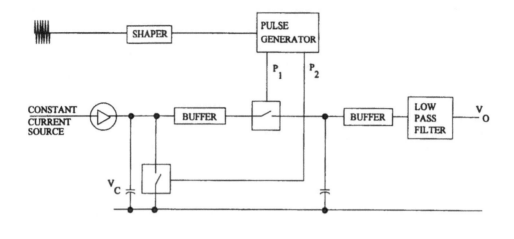

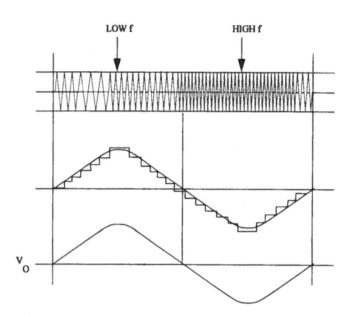

Figure 1-4 PPM receiver.

Figure 1-4 describes the action associated with the PPM receiver. A constant current source similar to that at the transmitting terminal charges a capacitor. The duration of the changing period is dictated by the frequency of the received signal. Pulse P_1 which is generated by the transitions of the squared incoming signal causes the voltage on V_C to be transferred (after filtering) to the output V_O. Pulse P_2 operates an electronic switch that shorts the capacitor, allowing a new cycle to start. The output V_O, as a result of this, is a reproduction of the waveform that is imposed on the transmitter. Proper consideration of transmission delay is included.

6. PHASE MODULATION

The sudden variation of phase angle is a third basic way in which a transmitter can provide information to a remote receiver. One obvious implementation consists of a polarity inversion. In Figure 1-5 this reversal is apparent, and the receiver is able to detect this easily. In this example, each reversal is identified as a change in state of a digital signal.

6.1 Quadrature Phase Shift Keying

In quadrature phase shift keying (QPSK), the shift in phase is implemented at specific angles, with the pertinent angle

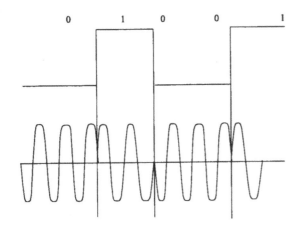

Figure 1-5 Phase modulation.

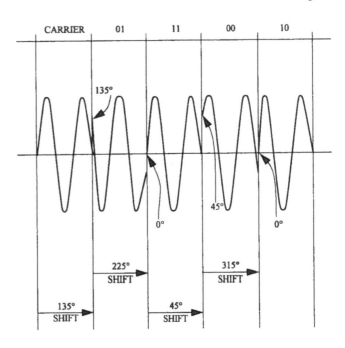

Figure 1-7 Example of the transmitted pattern for a QPSK system.

being the difference between the phase angle at the edge of the bit boundary compared with the phase shift at the edge of the previous bit boundary. If the shift at the edge of the previous bit were, for example, 90° and the present phase shift were 315°, this would be recognized as a 225° phase shift.

By relating a group of four phase shifts 45°, 135°, 225°, and 315°, four digits (two bits each) can be obtained: 00, 01, 10, and 11. The star diagram of Figure 1-6 defines this relationship. Because the difference angles correlating with these two bits patterns are at a 90° relationship, the keying mode is identified as a quadrature system. To send

an 8-bit pattern of 01110010 phase shifts of 135, 225, 45, and 315 would be initiated. Figure 1-7 describes the signal pattern that would result.

7. QUADRATURE–AMPLITUDE MODULATION

One final communications fundamental that has experienced important application in relaying is the quadrature–amplitude modulation (QAM) technique. As the name would lead one to expect, this system involves two different amplitudes for each phase shift. This allows four different states (4 bits) to be represented by each of the 16 different phase angle shifts and magnitudes. Figure 1-8 describes the bit pattern that is represented by each phase shift–magnitude combination. For example, if the phase shift at the edge of the bit boundary is 0° and the phase shift at the edge of the previous bit boundary is 135 degrees, this represents a 225° shift. If, at the same time, the level of received signal has been boosted to $3\sqrt{2}$ times the nominal level, Figure 1-8 shows that this represents a four-digit number (quad-bit) of 1110. The signal pattern associated with the transmission of the digital message is pictured at the top of Figure 1-9.

On being received, the digital signals are checked for validity and either used in digital comparisons or con-

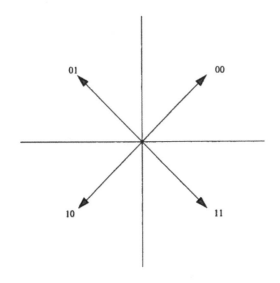

Figure 1-6 Constellation pattern for QPSK.

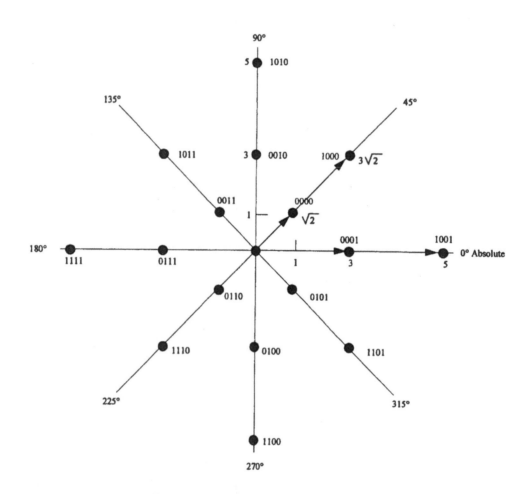

Figure 1-8 Constellation pattern for QAM.

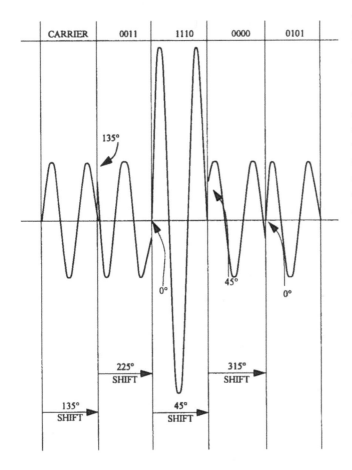

| CARRIER | 0011 | 1110 | 0000 | 0101 |

135°

0°

45°

0°

225°
SHIFT

315°
SHIFT

135°
SHIFT

45°
SHIFT

Figure 1-9 Typical signal pattern for QAM.

verted to analog signals for comparison at the receiving station.

7.1 General Comments

The obvious benefit from a choice of the more complicated forms of modulation is that more information can be delivered over a given channel. The on–off principle provides simply a 1 or 0 for use in the logic. FSK can provide three states, high, neutral, and low, but is generally operated two state. QPSK delivers 2 bits for each phase change. QAM accommodates 4 bits per phase shift and amplitude establishment. QAM can provide 9600 bits per second over a 2400-Hz channel.

2

Current Differential Relaying

WALTER A. ELMORE

1. INTRODUCTION

In all pilot relaying, a decision to trip or restrain is dependent on the availability of information from the remote transmission line terminal, as well as on local information. Current differential relaying is a method of extending the benefits of differential protection as applied to generators and transformers to the protection of transmission lines. The simplicity of comparing current flowing into a power system component to the current flowing out of that component is very attractive, with the important advantage of requiring no voltage transformers. It allows detection of internal faults, sensitively (limited only by the ability of the current transformers (ct's) to accurately reproduce the primary currents), while ignoring external effects, such as faults, load, and power swing conditions. Another important benefit is that simultaneous tripping at both transmission line terminals can be effected, irrespective of the relative current contributions from the sources at the two ends.

To extend these benefits to transmission line relaying requires faithful reproduction locally of some remote quantity by way of the communication channel. The quantities to be compared must be time-coincident (or adjusted), and the magnitude and angle of the remote quantity must be preserved. Two important mediums for allowing this comparison are pilot wires (two-conductor circuit capable of transmitting 60 Hz and dc quantities)

and fiber-optic cables (either dedicated or multiplexed). The nature of the system allows recognition of all types of phase and ground faults at high speed based on current comparison alone.

2. PILOT WIRE RELAYING

To limit the communications requirement to a single pair of conductors requires the generation of a single representative quantity, which falls within a reasonable range for all fault types. This is referred to as the filter output voltage. In the HCB and HCB-1 relays, advantage is taken of symmetrical components by using Eq. (2-1).

$$V_F = C_1 I_{A1} + C_2 I_{A2} + C_0 I_{A0} \tag{2-1}$$

where V_F = filter output voltage
I_{A1} = positive-sequence component of A-phase line current
I_{A2} = negative-sequence component of A-phase line current
I_{A0} = zero-sequence component of A-phase line current
$C_1 C_2$ and C_0 = weighting factors

By properly selecting the weighting factors, sensitive ground fault recognition is possible, while limiting the filter output voltage for large-magnitude phase faults.

9

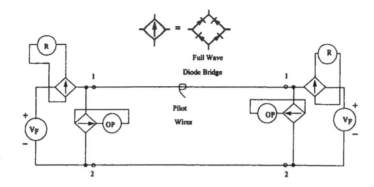

Figure 2-1   Simplified fundamental schematic of HCB-1 pilot wire relays.

V_F is developed at each transmission line terminal and compared over the pilot wires. The restraining coils are in series with the filter output, and the operating coils are in shunt, as shown in Figure 2-1. These two coils produce counteracting ampere-turns and cause torque to be produced by interaction with the polar unit electromagnet as seen in Figure 2-2. For an external fault, the two filter output voltages are essentially equal in magnitude and 180° out-of-phase. This causes a substantial circulating current to flow through the pilot wires and through the restraint coils. Only a small current flows through the operating coils at the two locations. The relays at both terminals are restrained from operating.

For an internal fault, the contributions to the fault will be substantially in-phase (although up to 90° phase difference can be tolerated), and they may have any combination of magnitudes. If the contributions to the internal fault from the two power system sources were equal and in-

phase, the two filter output voltages would be equal and in-phase. No current would flow in the pilot wires, and the operating and restraint coils would carry equal currents. Because the operating coil has many more turns than the restraint coil (see Fig. 2-2), a net operating torque results. Note that the absence of input current at one terminal (no contribution of current from one source) still allows high-speed tripping to take place at both terminals, as a result of current in the operating coil at both locations.

The foregoing description covers the basic operating principle involved in the HCB and HCB-1 relays, but proper control of energy level and isolation dictates the use of an internal saturating transformer and an external step-up insulating transformer (4:1 or 6:1). Figure 2-3 shows this in detail.

The filter output is generated in the HCB and HCB-1 relays with mutual reactors and resistors. These are not interchangeable relays, nor is one an extension or replacement of the other. The HCB has no output as a result of negative-sequence current flow, to assure identical outputs for all combinations of phase-to-phase faults, whereas the HCB-1 is designed to provide outputs, in power system terms, that are compatible with the fault currents for three-phase faults and phase-to-phase faults (i.e., $I_{\phi\phi} = 0.866\ I_{3\phi}$).

2.1 HCB Filter

Figure 2-4 shows the essential components of the HCB filter. The filter output voltage is described as V_F. This is a Thevenin voltage, the open circuit voltage as viewed from the output terminals.

One important aspect of symmetrical components is that their influence on apparatus may be examined, one component at a time, and the results combined to provide the same effect as the original phase quantities. This will be done for the HCB filter.

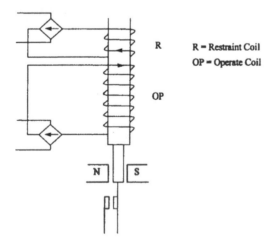

Figure 2-2   Interaction between operating and polarizing coils.

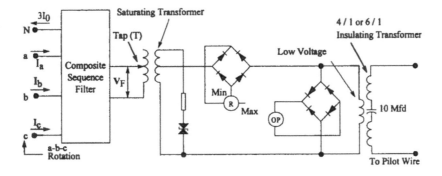

Figure 2-3 The ac connections to the HCB-1 relay.

For reasons, which will become apparent, consider first the application of negative-sequence current only. Because of the consideration of the open circuit at V_F, the A-phase current will flow through resistor R_1 in Figure 2-4. The reactor ab, in the loop abcd, will have a voltage induced in it that is proportional to the product of B-phase current and X_m, the mutual reactance between the B-phase reactor and the ab reactor. Owing to the reactive character of the mutual, the induced voltage drop leads the current producing it by 90.

Similarly, the C-phase current induces a voltage into reactor ab that is proportional to the product of the current and X_m, the mutual reactance. From observation of the polarity markings on the windings of Figure 2-4, it is evident that the B-phase effect is opposite that of the C-phase current. As connected, the voltage drop in the loop abcd is positive for the B-phase effect and negative for that of the C-phase.

With only negative-sequence current applied to the filter, the three-phase currents add to zero at point c, with no current flowing through resistor R_0. Thus, the net effect of negative sequence current flow is to produce a filter voltage of zero, as Figure 2-5 shows. Note that this occurs as

a result of the choice of $X_m = R_1/3$. R_1 and R_0 are settings and as R_1 is changed X_m must change also, maintaining the necessary ratio.

The filter is oriented toward the A-phase, but, under identical conditions, all combinations of phase-to-phase faults on the power system produce an identical *positive-sequence* current magnitude. Therefore, with the *negative-sequence* output being always zero, the filter output voltage will, for all phase-to-phase fault combinations, be the same for a given current level.

With only positive-sequence current applied, Figure 2-5 shows that the filter output voltage is $2I_{A1}R_1$. With only zero-sequence current applied, the filter output is $I_{A0}(R_1 + 3R_0)$ with the B- and C-phase mutual exactly canceling one another in reactor ab.

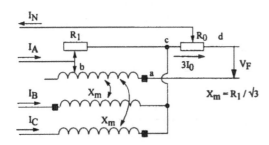

Figure 2-4 Simplified schematic of HCB filter.

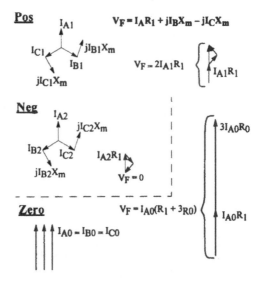

Figure 2-5 Phasors for HCB filter.

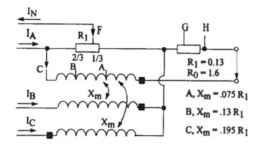

Figure 2-6 Simplified schematic of HCB-1 filter.

2.2 HCB-1 Filter

Figure 2-6 shows that similarities exist between the HCB-1 and the HCB (see Fig. 2-4) filters. However, note that the reactor polarities, the mutual impedance values, and the R_1 values are different. This leads to very different phasor diagrams and different filter output voltages (compare Fig. 2-5 with Fig. 2-7).

The HCB-1 filter produces a filter output voltage for a phase-to-phase fault (AB or CA) that is equal to the filter output voltage for a three-phase fault provided the $I_{\phi\phi} = 0.866\, I_{3\phi}$. This squares with the results the power system produces for a fault at any given location (except at a power plant where X_2 may be radically different from X_1). Because the orientation of the filter has A-phase as a reference, the sensitivity of the HCB-1 relay to BC phase-to-phase faults is only $0.53\, I_{3\phi}$.

The HCB relay has no response to negative-sequence current. Therefore the response to all phase-to-phase faults is identical. The filter output voltage is dependent entirely on positive-sequence current for all $\phi\phi$ faults, thereby requiring the same positive sequence current as for the 3ϕ faults for tripping to occur. Because nature does not provide the same positive sequence current, but provides only one-half as much for the $\phi\phi$ case as for the 3ϕ case, this produces no detrimental effects, but requires only attention in choosing the settings.

Both the HCB and HCB-1 have provisions for eliminating the zero-sequence response completely. Circulating zero-sequence current in parallel lines can be produced by positive-sequence fault current flow through unbalanced impedances (caused by the lack of transpositions). The very heavily weighted "through" zero-sequence current effect on the filter output voltage may obscure the effect of the internal fault positive-sequence current, thereby producing blocking action when tripping is the correct response. To avoid this, the weakening or elimination of zero-sequence response is required.

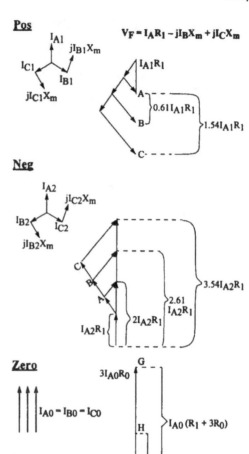

Figure 2-7 Phasors for HCB-1 filter.

The 2/3, 1/3 taps on the R_1 resistor facilitate the elimination of zero sequence effect on the filter output when required. Observing Figure 2-6, and considering only a zero-sequence current input, it is seen that cancellation of ampere-turns occurs in the three-winding reactor. I_0 flows left-to-right through 2/3 R_1. Similarly (on tap F) $2I_0$ flows right-to-left in 1/3 R_1. With these two voltages being produced, which are equal in magnitude, but opposite in direction, there is, then, a net voltage of zero resulting from the zero-sequence current input, when tap F is selected.

3. PILOT WIRE PROTECTION

With the power system substations being widely separated, the pilot wires between the two may be subjected to

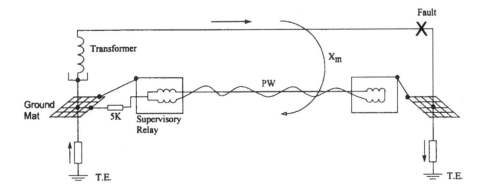

Figure 2-8 Source of extraneous voltage in pilot wire circuits.

the introduction of extraneous voltages. Zero-sequence mutual effects between nearby power system circuits and the pilot wires interconnecting the pilot wire relays may be severe. These voltages are end-to-end on the pilot pair and may be nullified by mutual drainage reactors inserted in shunt that allow compensating current to flow through the pilot wires. The other possible source of extreme voltage that may affect the pilot-wire–relaying system is difference in ground-mat voltage. This voltage may be accommodated by the introduction of neutralizing reactors in series with the pilot pair.

Figure 2-3 shows the manner in which the pilot wires are connected to, but isolated from, the pilot wire relays. The supervisory relays, described later, are connected across the 10-microfarad (μF) capacitors. Their cases are tied to station ground-mat. The "sending-end" supervisory relays also contain elements that are connected to station ground-mat. Thus, these extreme voltages may be imposed between the receiving-end relay internal components and its case, which is connected to *remote* station ground. These relays (and generally the pilot wires) are capable of supporting 1500 V rms (root-mean-square) for a short time, but the extraneous voltage may exceed this appreciably.

Figure 2-8 describes, in simple terms, the process by which these extraneous voltages may be injected. A grounded-wye transformer is shown contributing to a phase-to-ground fault in a adjacent station. The ground-mat voltage at one station rises while the ground-mat voltage at the other drops, thereby producing a difference of, perhaps, several kilovolts (kV) depending on fault current magnitude and the resistance from each of the station grounds to true earth. The same fault current, in this example, produces a flux that links the pilot wires and induces a voltage that is dependent on fault current magnitude ($3I_0$), distance between the pilot wires and the power

conductor, and length of exposure. Typical voltage profiles are shown in Figure 2-9.

Two devices that are typically applied to nullify these effects are described schematically in Figure 2-10. The neutralizing reactor, as shown, is a two-winding device having polarities that allow the pilot wires to be held at remote ground potential, whereas the supervisory relays go to the voltage level dictated by the ground-mat rise (or drop). The mutual drainage reactor with its gas tube or voltage protector allows current to flow from end to end, equally in the two wires. This current level and direction is such that it tends to nullify the original flux linking the pair.

Figure 2-11 emphasizes the advantage of using neutralizing reactors to overcome the effects of differences in sta-

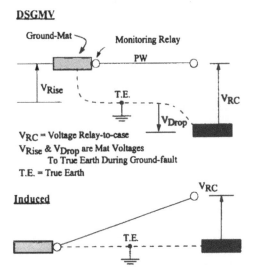

Figure 2-9 Voltage profile considering influences one at a time.

tion ground-mat voltage (DSGMV). A device that concentrates the voltage correction for a concentrated voltage effect is superior to one that distributes the voltage correction. *The sheath or shield should in no case be connected to the station ground-mat when neutralizing reactors are used.*

It is extremely important to recognize the safety hazard associated with the use of neutralizing reactors. During ground faults several kilovolts may exist across the neutralizing reactor and, therefore, between the pilot wires and station ground. The terminals of the pilot wires connected to the neutralizing reactor must be screened or physically isolated in some way to prevent any possible personnel contact between them and apparatus or panels connected to station ground.

Figure 2-12 shows the superiority of using the mutual drainage reactor to dissipate, in a distributed way, the distributed effect of mutual reactance. The relays are held to station ground voltage level and the pilot conductor to shield voltage is held to a very small value using mutual drainage reactors. Neutralizing reactors protect the relays, but the conductor to shield voltage may be large. Finally, Figure 2-13 describes the use of both devices for protecting the pilot wires. This apparatus is required at both stations.

In general, mutual drainage reactors are required where monitoring relays are used, and the induced voltage (end-to-end) may exceed 1200 V. The L2A600 voltage protec-

tors breakdown at approximately 400-V–ac rms and can tolerate 200-A rms for 11 cycles. The pilot wire resistance is the principal limiter of current flow. Neutralizing reactors are required where monitoring relays are used and the difference in station ground-mat voltage can exceed 1200 V.

The voltage protectors in this figure (because of the use of neutralizing reactors) must be connected to remote ground with a conductor that is isolated from station ground. *Remote ground* is defined as that location at which potential is not influenced by the change in voltage of the substation ground mat. A distance of several hundred feet from the station is adequate for establishing the ground, and an insulated conductor compatible with the short-time capability of the voltage protectors is adequate.

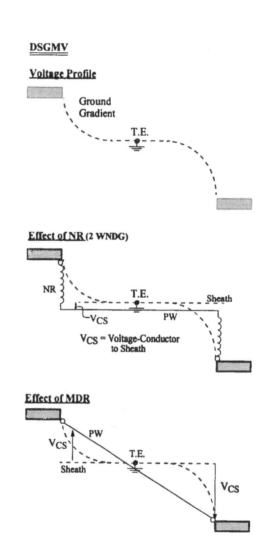

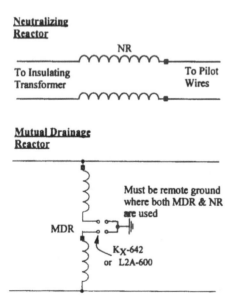

Figure 2-10 Protective apparatus for use with pilot wire relays.

Figure 2-11 The effect of the two devices of Figure 2-10 in nullifying differences in ground-at-voltage.

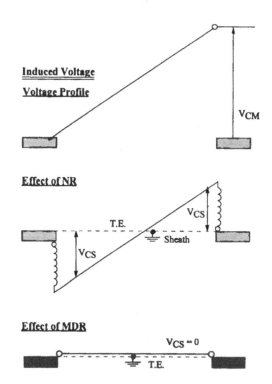

Figure 2-12 The effect of the two devices of Figure 2-10 nullifying induced voltage.

The scheme of Figure 2-14 may be used for very short line applications, where remote ground cannot be established conveniently. It uses the remote station ground-mat as the reference, rather than using true-earth. The mutual drainage reactors and protective gaps are required only if there is a "mutual" problem. The large conductor (shown as 4/0 CU) is required to protect the shield in the presence of large current flow resulting from mutually induced voltage.

4. CURRENT DIFFERENTIAL RELAYING

By using operational amplifier circuits for phase shifting and summing, the same type of filter output voltage is developed in the LCB-II as for the HCB and HCB-1 relays. To overcome the susceptibility to the extraneous voltages that pilot wires can impose, the LCB-II uses audiotones or fiber-optic signals.

To provide the ability for direct comparison of the filter output voltages as in the HCB and HCB-1 relay, a communication scheme is used (in both directions) that re-creates at the far terminal an "instantaneous" voltage that matches the voltage at the near terminal. The relay waveform is transported to the remote terminal, as described in Chapter 1 Section 5, Pulse-Period Modulation.

The "far" and "near" values are then compared. A restraint voltage is generated that is the sum of the magnitudes of the local and remote filter output voltages. An operating voltage is generated that is the magnitude of the phasor sum of the local and remote filter output voltages, this being proportional to the differential current. Tripping occurs when the operating voltage exceeds 0.7 of the restraint voltage. Summarizing:

$$V_R = |V_L| + |V_R| \quad restraint$$
$$V_0 = |V_L + V_R| \quad operating$$
$$V_{trip} = V_0 - 0.7 \, V_R > V_{PU} \quad trip$$

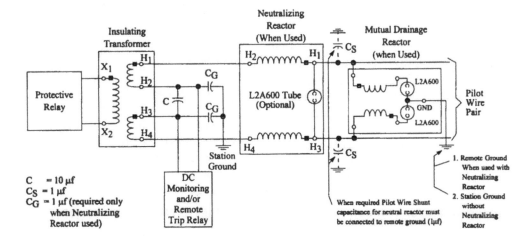

Figure 2-13 Neutralizing reactors and mutual drainage reactors used for protection of pilot wires.

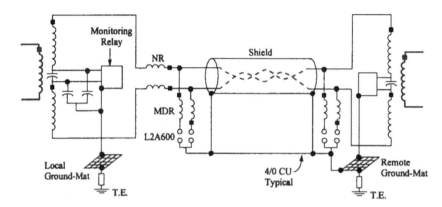

Figure 2-14 Alternative arrangement using remote ground-mat for reference, rather than true earth (note: full ground-mat voltage difference must be supported by one neutralizing reactor).

The 0.7 factor was selected to assure the ability to trip for an internal fault even though the sources at the two ends of the protected line are as much as 90° apart. V_{PU} is a level detector setting.

As with the HCB and HCB-1, this relay is able to detect sensitively, and at high speed, all types of phase and ground faults on the protected line, as well as being able to do so with no fault current contribution from one terminal.

The fiber-optic and tone systems inherently have delay in the communication system. Provision must be included to assure proper alignment between the local and remote quantities for an external fault. Because nothing can be done to advance the remote quantity, then the expedient that is exercised is to delay the local quantity to match the channel delay.

4.1 FCB

The FCB is a converter box that allows the HCB or HCB-1 relay to work into a dedicated fiber-optic circuit. Two fibers are required, one for each direction of transmission. No modification of the HCB or HCB-1 relay is required for this application. It eliminates the need for insulating transformers or any protective apparatus, such as is applied to pilot wires. Supervision of the channel is accomplished by the FCB itself. Although this is a very secure scheme, it does not make the most efficient use of the fiber.

4.2 Microprocessor Implementation

A very important recent addition to this list of current differential relaying methods utilizes microprocessor technol-

ogy (REL 356). The symmetrical component filter takes advantage of a broad range of engineering tools. The individual currents, I_A, I_B, and I_C, are converted to Clarke components. The nature of symmetrical components is that the quantity a ($-0.5 + j0.866$) involves a phase shift. Conversion of a set of phase currents to Clarke components does not, for only "real" multipliers are involved. The further conversion to symmetrical components (more respectfully known as Fortescue components) is also accomplished without the need for phase shift. Elimination of phase-shifting multipliers greatly simplifies the protection algorithm. This version of current differential relaying uses the traditional equation for combining and comparing:

$$V_F = C_1 I_{A1} + C_2 I_{A2} + C_0 I_{A0}$$

Samples of V_F are taken and transmitted, using a tone channel or fiber optics, for use in comparing phasor quantities at the remote terminal. A similar comparison is made locally of the local V_F and the received V_F from the far terminal.

4.3 T1 Carrier

Systems are available that use time-division multiplexing to accommodate 24 channels over one fiber (in each direction), with one channel being allocated to the current differential function.

5. PILOT WIRE SUPERVISION

One successful scheme that has been used widely to supervise the integrity of the continuous metallic pilot wires that are used in the HCB or HCB-1 scheme involves the introduction of 1 mA of dc circulating current. A supervisory relay detects the presence of this current. As viewed from the "sending" end, low current represents an open conductor, and high current represents a short circuit. At the "receiving" end, an open or short causes this current to be low, allowing detection at either or both terminals of difficulties in the pilot wire circuit.

A remote trip function is also available through the use of a reversal of direction and increase in the magnitude of the circulating current. A separate remote trip element in the receiving relay recognizes this reversal of direction and operates to produce tripping at a terminal remote from the initiating terminal.

6. THREE-TERMINAL APPLICATIONS

All of these variations of pilot wire and current differential relaying can accommodate three-terminal applications. Each terminal must be equipped with identical protective relaying with each connected identically for pilot-wire relaying. External faults produce circulating current (ac) over the pilot wires that creates restraint in the relays at all three locations. Internal faults produce filter output voltages that cause little circulating current (none if the contributions from all power system sources are identical) and predominant operating energy at all three terminals, even though a small outfeed may exist at one terminal for this internal fault.

For the LCB-II current differential scheme, each filter output voltage is recreated at each of the other two terminals. The operating quantity is, as for the two terminal applications, equal to the magnitude of the sum of the three quantities. The restraint is, again, equal to the sum of the magnitudes. Operation occurs when the operating quantity exceeds 0.7 of the restraint value.

7. SUMMARY

Current differential relays allow the benefits of conventional differential relaying to be achieved, even though the current transformers that define the boundary of the protection zone may be many miles apart at the terminals of a transmission line. Various communications media have been used with great success for this relaying system, from pilot wires to fiber optics. All types of faults are located and isolated at high speed.

3

Pilot Channels for Protective Relaying

WALTER A. ELMORE

1. INTRODUCTION

A communications channel is an integral part of a pilot relaying system that is used to protect a transmission line. It is needed to deliver information from one terminal of the line to the other, so that a fault location can be determined. Similar information is also conveyed in the opposite direction, allowing essentially simultaneous determination of the presence of an internal fault.

The requirements of relaying communications are much more stringent than for data or voice transmission. Temporary blanking of information can be retrieved or repeated, but no such possibility exists for relaying. During the very brief interval of a fault, a clear unambiguous interpretation of the state of the channel must be available. The state of the channel, in turn, must be dictated by the protective relays. An incorrect interpretation will lead to an undesired trip or a failure to trip.

2. TYPES OF CHANNELS

The predominant communications channels that are used in protective relaying are as follows:
A. Power line carrier
    1. On–off
    2. Frequency-shift
    3. Single–side-band

B. Audio tones
C. Microwave
    1. Analog
    2. Digital
D. Fiber optics
E. Pilot wire

3. CHOICE OF CHANNELS

Figure 3-1 provides a rough indication of the relative cost of several types of channels. The pilot wire channel cost increases linearly with distance, but is unquestionably the lowest-cost channel for short distances. A single microwave channel is hardly justifiable, but as the diagram shows, the incremental cost for additional channels is small. Because microwave is a line-of-sight medium, and terrain and earth contours limit the distance over which transmission can take place reliably without repeat terminals, there is a jump in cost at the limiting distance.

Power line carrier is not limited by distance, in general, but the terminal cost is appreciable. Line traps, tuners, and coupling capacitors are required. The line traps must be able to conduct the maximum transmission line current without excessive heating, have the appropriate inductance and capacitance, and be able to support the forces associated with the maximum short-circuit current. It is supported in the line, and requires no independent insulating

19

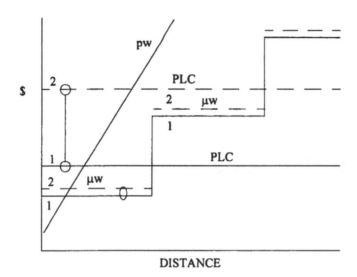

Figure 3-1 Relative costs of different channel types.

structure. The coupling capacitor must have appropriate capacitance for coupling the carrier signal onto the high-voltage circuit and be fully insulated to support the steady-state and transient voltages to which the power system may be subjected. There is generally a gap-to-ground provided in the tuner, and in the base of the coupling capacitor, that will spark-over in the event excessive voltage appears across the lead-in cable between these two devices. This voltage can result from lightning, switching surges, or faults. Vigilance should be exercised in maintaining proper gap spacing because flashing of the gap causes shorting of the carrier signal, and this may occur at the instant at which the carrier signal is needed most.

A dedicated power line carrier utilizes little that is common between multiple channels. Tuners and line traps are more costly, but the coupling capacitor need not be duplicated. Apparatus is necessary to avoid intermodulation products from interfering with normal transmission of the signal. This increased cost for multiple channels is shown in Figure 3-1.

Two other significant communication channels that are not shown in the figure are single–side-band carrier and fiber optics. Single–side-band carrier accommodates multiple channels in a manner comparable with that in microwave, but has a substantially higher terminal cost than a dedicated power line carrier. Fiber optics is similar to pilot wire costs, in increasing linearly, but the cost per mile is *considerably* higher.

Dedicated fiber for a single function is occasionally used, but this is an inefficient use of an expensive resource. A very wide-frequency spectrum is available through the use of fiber optics, and it is customary to utilize it more fully with multiple channels. This arrangement involves the use of pulse code modulation at the T1 level and accommodates 24 channels. Only one of these channels is required in each direction for a pilot-relaying system.

4. POWER LINE CARRIER

The references at the end of this chapter are replete with extremely useful information on the application of power line carriers. These references should be used for detailed analysis of a particular power line carrier application. However, Figure 3-2 shows a representative pilot-relaying application with typical attenuation values.

4.1 Modal Considerations

In much the same way as for symmetrical components, individual, unbalanced phase currents may be resolved into sets of balanced currents each of which behave independently without interaction. These may then be recombined after analysis to regenerate the individual phase currents. Power line carrier modes most closely resemble Clarke components. It must be recognized that the behavior of high-frequency (kilohertz; kHz) signals are quite different from power frequency quantities, and this accounts for the differences in the concepts. Figure 3-3 describes the three sets of modes that are pertinent to power-line carrier. Because of the large attenuation associated with modes 2 and 3, the prevalent mechanism by which power-line carrier is

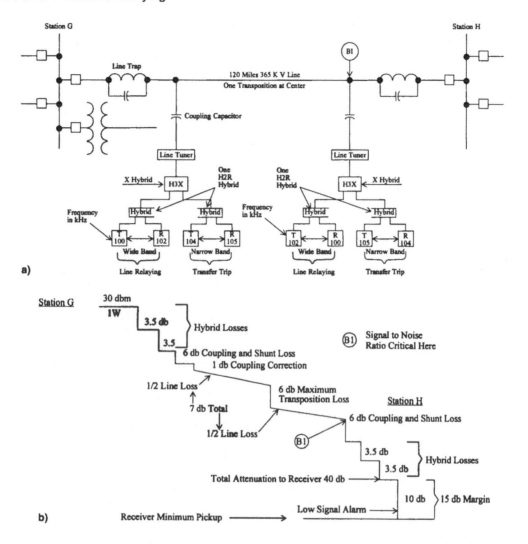

Figure 3-2 (a) Schematic diagram of a typical carrier system; (b) loss profile for a from G to H.

transmitted is mode 1. Coupling a single signal to the center phase of a transmission line through a tuner (reactor) and a coupling capacitor introduces both mode 1 and mode 3 signals.

For purposes of understanding the phenomenon, p in Figure 3-3 can be assumed to be 2 and q can be set to equal 1.0. The combination of mode 1 and mode 3 which satisfies the condition of introducing carrier power into the center phase is with mode 1 = 2/3 and mode 3 = 1/3. To relate mode 1 power to the power introduced in the phase, we may compare $(1/3)^2 + (2/3)^2 + (1/3)^2$ with 1. This gives mode 1 power = 6/9 of the power introduced. Converting this to decibels (dB), we get 10 log 2/3 = −1.76 mode 1 coupling efficiency. Mode 3 coupling efficiency may be determined as 10 log 1/3 = −4.77 dB.

More careful appraisal of p and q for a typical 500-kV line is shown [1] to yield −1.6 dB and −5.12 for mode 1 and mode 3 loss, respectively for center phase-to-ground coupling.

4.2 Typical Loss Profile

Each component of the transmission path must be evaluated for its effect on the total attenuation between transmitter and receiver. Figure 3-2a portrays a typical power-line carrier system and Figure 3-2b shows a profile of the losses associated with the left-to-right transmission. A similar, but not necessarily identical, profile may be developed for right-to-left signal transmission.

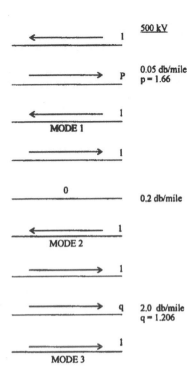

Figure 3-3 Power line carrier modes.

4.3 Use of Hybrids

If hybrids are used to permit multiple channels to cooperatively use the same coupling devices for various functions, their losses must be evaluated. Usually 3.5 dB will be the value for resistance hybrids with 0.5 dB for skewed hybrids for the transmitter and 3.5 dB for received signals.

4.4 Coaxial Cable

Coaxial cable may be evaluated as 0.45 dB/1000 ft at 30 kHz and 0.92 dB/1000 ft at 300 kHz for RG-8U cable, with essential linearity for frequencies in between: larger cable produces lower loss, and smaller cable produces more.

4.5 Tuners

Tuners are required to provide a series tune with the coupling capacitor at the carrier frequency. Where only a discrete dedicated frequency is involved, a sharply tuned, single-frequency tuner may be used, producing as little as 1 dB loss at the desired frequency. Broader-band requirements, when multiple channels are in use, will dictate the use of a broadband tuner which, in turn, will manifest a higher loss (for each of the channels).

4.6 Noise

Noise is a key ingredient in the determination of the degree of success of a power line carrier system. Noise is generated by corona, lightning, and other communications channels, with corona producing white noise (energy uniformly distributed throughout the frequency spectrum), and lightning producing impulse noise. The signal-to-noise (S/N) ratio must be above a certain level to assure proper distinction between the signal and the noise. The wider the bandwidth of the receiver the greater the susceptibility to white noise. The generally used figures of required in-band S/N ratio are 15 dB for on–off carrier and 13 dB for frequency-shift carrier. Most of the field investigations of carrier noise have involved the use of measuring apparatus with a 3-kHz bandwidth.

The signal-to-noise requirement for receivers obviously is based on in-band noise. The "standard" noise values collected through field studies must be modified to conform to the actual bandwidth in hertz as compared with the 3-kHz data. The power admitted to the actual receiver is bandwidth/3000. With this knowledge, it is evident that a much higher level of out-of-band noise can be tolerated by a narrow-band (220 Hz) frequency-shift receiver than by a wider-band (1500 Hz) on–off carrier receiver. Another influence of bandwidth is speed. In general, the more narrow the bandwidth, the longer the time delay associated with the filter. Several variations of bandwidth (and thus S/N ratio limits and speed) are available in commercial power-line carrier.

5. AUDIO TONES

As the name implies, audio tones fall within the frequency band that is audible to the human ear. These low-frequency channels are used

1. Independently on a pair of pilot wires
2. On a voice channel on power line carrier
3. On a voice channel on microwave

They predominantly fall into the frequency-shift category, as described in Chapter 1. In general, the tone is operated at its guard frequency and is shifted to its trip frequency when some action is required at a remote location. One important use of tones in a relaying context is for a transfer-trip function, as associated with the need to operate a distant breaker to clear a fault or to isolate a faulty breaker. They are also used in transmission line pilot-relaying schemes.

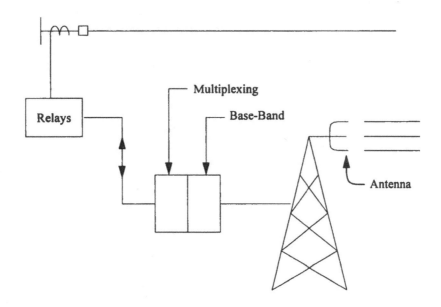

Figure 3-4 Typical microwave configuration.

6. MICROWAVE

The region from 1000 to 30 GHz is considered the microwave region. Microwave transmission involves a line-of-sight path. It is capable of transmission for only about 20–50 miles, depending on terrain, and is influenced for long hops by the curvature of the earth. Similar to light, microwave energy can be focused. It uses parabolic antennas that allow focusing of the signal power. The intent of microwave is that it communicates with only a single receiving antenna. When the distance exceeds the allowable value, repeaters may be introduced.

6.1 Basic Microwave Configuration

Figure 3-4 describes the basic microwave equipment configuration for a relaying application. The low incremental cost for additional channels will invariably involve joint use for the relaying with other functions, such as voice, supervisory control, telemetering, and status identification. The signals to and from the relays may or may not feed through the multiplexer. A higher level of dependability is achieved for the very important relaying function by working directly into the baseband equipment.

6.2 Microwave Reflection

Figure 3-5 describes an interesting microwave application that takes advantage of the reflective quality of the medium. A large billboard type structure is used to deflect the

microwave signals to the desired path. This would, for example, allow a hydroplant in a valley to be able to communicate to a remote substation with whatever functions are required. The reflector terminal is entirely passive, providing no signal amplification whatever.

Transmission of a microwave signal beyond roughly 35 miles, requires a repeater terminal. This is simply a receive, amplify, and retransmit terminal.

6.3 Multiplexing

The multiplexing is one of two types: frequency division or time division. Frequency division (referred to as analog) is a continuous process in which each channel occupies a different frequency and is identifiable on that basis. Time division multiplexing (referred to as digital) consists of interleaving the various channels by sampling them one at a time, in a specific pattern. Pulse code modulation (PCM) is the modern version of time division multiplexing. It combines a collection of functions into a continuous serial signal stream for application to a single communications path.

There are ten or more different coding systems in the PCM category that have been proposed and are used in different applications. Each involves a different clever interpretation of the simple transition from a 1 to a 0 and a 0 to a 1; each developed in the interests of obtaining a greater data rate on a channel having a given bandwidth.

The QAM system (described in Chap. 1, Sec. 7) takes this a step farther by incorporating multiple bits into each

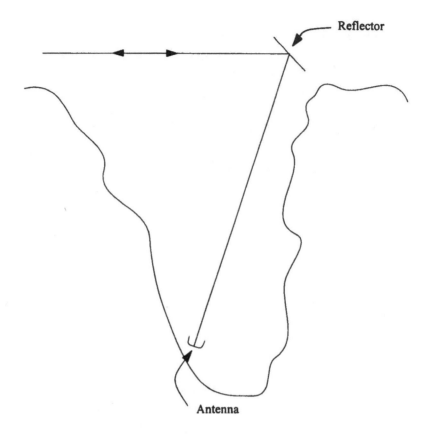

Figure 3-5 Use of a reflector.

transmitted state. Eight different phase shifts and two different magnitudes allow a representation of 16 different groups. This allows 4 bits to be identified with each signal, rather than 1. A 2400-baud channel, therefore, can accommodate 9600 bits/s (bps).

Errors occur in data transmission, and various means are used to detect or correct them to assure that action dictated by the erroneous signal is not carried out. Parity check is effective, but multiple errors may go undetected. Checksum is another expedient that is exercised by adding together a group of digital "words" at the transmitting location and transmitting the sum as part of the overall message. Comparison at the receiving location of the two sums confirms the validity of the data. Several different methods of error detection are often applied to important data streams.

7. FIBER-OPTICS

Optical communications allow massive amounts of data to be transmitted from one location to another in an environment that is free of noise or interfering signals. The fiber-optic cable provides the medium through which the light signals pass. The noise in the channel is generated within the light detector and the amplifiers at the receiving station where the designer can deal with it as required.

There are three different types of fibers: stepped-index multimode, graded-index multimode, and stepped-index single-mode. Their classification is based on their construction and the refraction qualities of the fiber and the cladding. Their costs, bandwidth, and attenuation factors vary widely, and expert advice should be obtained for their selection and application.

Any of these fibers are suitable for relaying use, provided the distances (total attenuation) are within the limits for the particular fiber chosen. A fiber-optic channel is capable of supporting a much higher data rate than traditional relaying systems require. Rather than dedicating a fiber to relaying alone, it is often incorporated into a multiplexing system in which the fiber is shared with other functions (*any* functions). One such system is called T1, having the capability for 24 voice channels, each operating at 64 kbps. The same wide varieties of multiplexing systems, as described under microwave, are also used with fiber-optic transmission.

In many relaying applications, an analog quantity (continuous, uninterrupted sinusoid) is needed at a remote location for comparison with a comparable analog quantity developed at that remote location. This is done by sampling the instantaneous value of the sinusoid at a rate determined by the degree of resolution required and the practical availability of the hardware, forming a binary code representing that magnitude, and transmitting the digital signal over the fiber channel to be reconstituted back into analog form with appropriate filtering at the remote location.

8. PILOT WIRES

Metallic conductors have been widely and successfully used over the years with current differential relays. This was described in Chapter 2. Although completely applicable for short-line relaying, their limitations are impressive. Their transmission frequency is 60 Hz, disallowing any possibility to filter out the noise. Under conventional supervision, the circuit must pass dc, and is subject to huge common mode voltages for cases (external faults) that they must ignore. In spite of these difficulties, virtually every utility and major industrial facility on the continent have used and are using current differential relaying on pilot wires.

9. ADDITIONAL INFORMATION

Reference 2 contains a complete description of power-line carrier channels, and Reference 3 covers the application of fiber optics to protective relaying and other utility communications functions.

REFERENCES

1. MC Perz. Natural modes of power-line carrier on horizontal three-phase lines. IEEE Power Apparatus and Systems. 1964, pp. 679–686.
2. RE Ray. Channel considerations for power-line carrier. ABB Relay Product Leaflet 83-3.
3. RE Ray. Fiber-optic communications for utility systems. Georgia Tech Relaying Conference Proceedings, April 1993.

4

Transmission Line Pilot Relaying

WALTER A. ELMORE

1. INTRODUCTION

Pilot relaying is characterized in that it cooperates with a communication channel to identify the conditions that exist locally for a remote transmission line terminal. A non–pilot-relaying system may be used, for all except very short lines, to provide high-speed tripping for one or both of the line terminal breakers, depending on fault location. Pilot relaying assures the ability to trip *both* terminals at high speed for *all* faults on the protected circuit. The benefits of this are

1. Decreased fault damage
2. Improved power system stability
3. Decreased effect on nearby generation and load

2. BASIC PILOT-RELAYING SYSTEMS IN USE

Certain fundamentally different schemes have emerged, all of which yield satisfactory results, but each of which has certain appealing qualities and subtle shortcomings. The goal of all protection is to examine the currents entering a piece of apparatus and to compare these with the currents leaving it. For generator protection, for example, this is easily accomplished because the currents are accessible, and the current sources and the relays are in reasonably close proximity to one another. The length of a transmission line, however, prohibits this simple approach.

The earliest efforts to take advantage of the differential principle led to the "pilot–wire"-relaying concept described in Chapter 2. To limit the wiring interconnecting the two transmission line terminals, the concept of developing a single-phase voltage proportional to the symmetrical component content of the input currents was born. By comparing the phasor relation between this single-phase voltage at one transmission line terminal with that at the other(s), an internal fault can be differentiated from an external one. When using a wire-line or fiber-optic channel, this is still a very effective method for protecting a fairly short transmission line.

To extend the benefits of this form of relaying to longer transmission lines, the phase-comparison concept was conceived. This scheme compares the phase angle relation between these single-phase voltages with the aid of the communications channel.

Another widely used concept falls under the broad category of directional comparison. In its rudimentary implementation, directional phase and ground units sense the direction to a fault. If these units, at both transmission line terminals, agree that the direction to a fault is toward the protected transmission line, then tripping is required at both terminals. If either disagrees, the fault is construed to

27

be external to the protected line, and no tripping by that relaying system is desired.

3. PHASE-COMPARISON RELAYING

Information on the phase angle of the single-phase filter output voltage, described previously, is accomplished with an on–off system, or with a frequency-shift system.

3.1 On–Off Channel

Two obvious choices existed in the design of this system: to have the trip interval be the on period or to be the off period. Using power line carrier, the off period was the logical selection because the protected power line is the invariable path for the carrier signal. To avoid the (rare) possibility that a short circuit on the line might also be a short circuit to the carrier signal, the off period is distinctly suited for tripping. Adequate carrier trapping prevents external faults from shorting the carrier signal.

With the carrier off period being the tripping interval, the carrier must be keyed on during one-half cycle of the single-phase filter output. This was chosen to be the positive half-cycle. See Figures 4-1 and 4-2 for typical waveform patterns for internal and external faults.

Single-frequency operation is used with on–off carrier (i.e., the transmitting frequency at all terminals is the same), and all receivers respond to all transmissions. Tripping at any location can occur only if all carrier is off, which is to say the filter output voltage is negative at all locations, and that this condition must persist for a long enough period to assure security (3 or 4 ms).

To assure coordination between locations for an external fault and to accommodate that there is a finite transmission time and receiver buildup time, a phase delay is required. For time coincident comparison, then, the local signal must be delayed by an amount equal to this channel delay.

These systems are equipped with fault detectors to provide another level of security against false operation. To make certain, for an external fault, that carrier may be turned on as required for blocking tripping during the appropriate half-cycle, a second level detector is used with a lower setting. Being set lower and dealing with essentially the same current values at each end of a two-terminal application, the carrier can always be keyed for any external fault location for which a fault detector can operate. Chapter 5 on three-terminal applications describes the adjustments required in the settings to accommodate the third terminal.

When using an on–off carrier, tripping speed is fault-incidence–angle-dependent. That is, if a fault occurs at the most unfavorable time, even with exact trip alignment of the local and remote square waves, the coincidence timer may time-out for say 3.9 ms (when set for 4.0 ms) and then have to wait 8 ms during the blocking half-cycle, then time for the full 4 ms on the next half-cycle before initiating trip. On the other hand, a fault occurring at the most favorable moment can identify the need to trip in 4 ms. There is, then, a variation of tripping time of 12 ms, depending on the fault-incidence angle. In spite of this, tripping can be accomplished with this system in 12–24 ms.

To minimize the dependence on fault-incidence angle, a frequency-shift channel may be used.

3.2 Frequency-Shift Channel

As the name implies this type of channel (power line carrier or tones) operates at two different frequencies, depending on the input to the transmitter. When at rest with no input from the relaying equipment, a "space" frequency is transmitted. When an input above a preset level is applied, the frequency is shifted to a different level (the shift being 200 Hz, or thereabouts). The receiver recognizes this shift in frequency by the transmitter as being a different state, called "mark." These variations between space and mark are in response to the variations of the single-phase quantity, described previously, which is a composite sinusoidal value that is dependent on the weighted symmetrical components of the three-phase input currents.

If one then has available at both transmission line terminals a "local" signal (the sinusoidal value converted to a square wave) and the received mark and space information, two comparisons can be made to identify their relationship and thereby, fault location. Local 1 is compared with mark and local 2 is compared with space. If local 1 and mark are coincident, it is evident that the symmetrical component value at both terminals is positive and there is an internal fault. If they are in opposition, the two terminals have currents that are noncoincident and identify an external condition (fault, load, power swing). If the local 1–mark or local 2–space comparisons identify an internal fault, tripping is initiated.

With a frequency-shift channel, continuous keying and continuous channel supervision may be used, adding another benefit to this mode of operation.

3.3 Segregated-Phase Comparison

Through a communications channel, these systems provide a comparison of local quantities (which are representative of each phase current) and the quantities associated

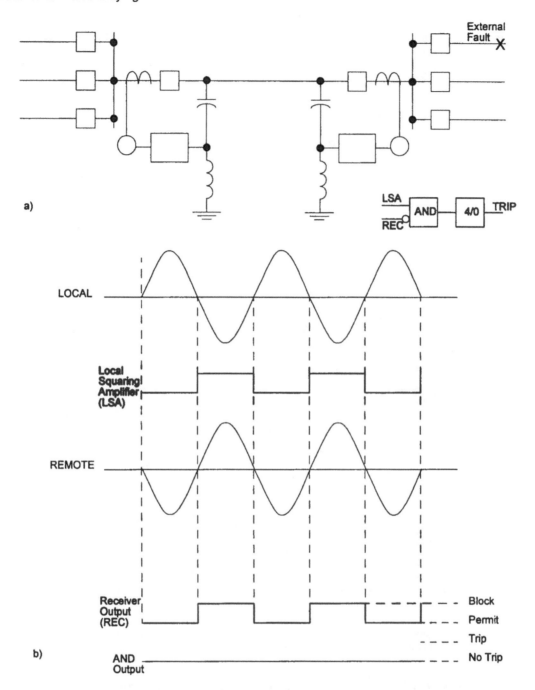

Figure 4-1 Phase comparison using on–off power line carrier (external fault).

with the identical phase currents at the remote transmission line terminal. From this comparison, an external fault may be identified by the fact that the current alignment is essentially 180° out-of-phase, irrespective of the type of fault.

For internal faults, the alignment on an individual-phase basis is more nearly in-phase, but not necessarily exactly in-phase, considering that the sources at the two ends of the line may be substantially out-of-phase, and also, the system parameters may be dissimilar at the two ends of the line.

These systems may be equipped with a ground current comparison system in the interests of providing more sensitivity to ground faults. In earlier segregated-phase com-

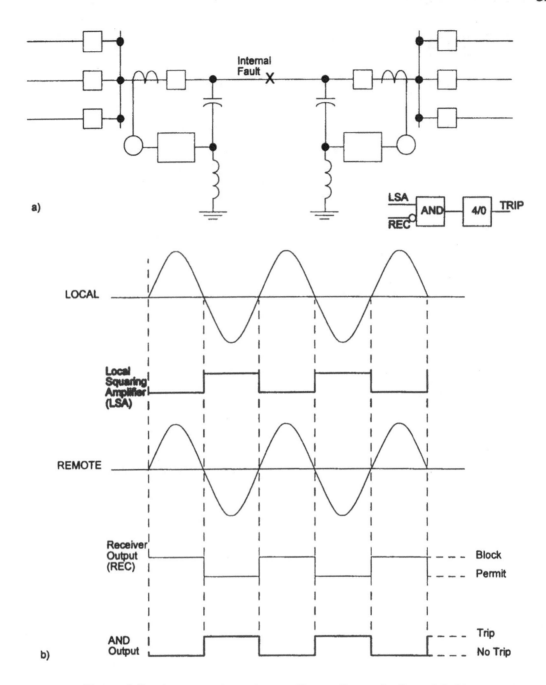

Figure 4-2 Phase comparison using on–off power line carrier (internal fault).

parison systems, this complicated the communications requirements because each subsystem must identify its state at the remote terminal (positive-going or negative-going). The earlier systems required essentially four channels each way for a three-phase and ground system.

To reduce the communications requirement a delta-ground system was used, consisting of a comparison over a single channel of I_A–I_B (delta I) each way, and a separate

comparison of $3I_0$. This required only two voice channels each way, but required careful evaluation of maximum load in establishing its settings.

Modern communications methods have allowed more information to be transmitted over a given bandwidth. This has allowed phase comparison to be accomplished for each phase and ground current at the two ends of a transmission line, with only a moderate increase in cost and complexity over that required for conventional single-quantity phase comparison.

In phase comparison, the key information that must be identified for tripping to occur is whether the currents from both terminals are simultaneously positive (i.e., both flowing into the protected line) or simultaneously negative with appropriate safeguards against such things as load and line-charging current. To identify a positive current condition for three phases and ground, simply requires the transmission of 4 bits, a 1 for positive current and a 0 for negative current for each of the four currents. A local quantity representing current compared with a received remote quantity, which has an opposite polarity, identifies an external load or fault condition and the need to refrain from tripping.

A microprocessor-based segregated-phase comparison system, utilizing a quadrature–amplitude modulation system is able to accommodate the communications requirements for four subsystems (three-phase and ground) over a single voice channel (each way). Faster speeds and more sophisticated logic are accomplished using a wider-band channel such as a fiber-optic one.

This concept has the unique advantage of being ideally suited for single-pole tripping, or for protecting transmission lines equipped with series capacitors. Faults involving more than one transmission line impose no difficulty whatever, whereas distance relay-based systems may become disoriented with either single-pole tripping, series capacitors, or multicircuit faults (called cross-country faults).

3.4 Evaluation of Phase-Comparison Systems

Phase-comparison relaying has the distinct advantage of immunity to zero-sequence mutual effects. It also, inherently, is not susceptible to incorrect operation because of power system swings or instability.

The segregated-phase comparison system is ideal for single-pole tripping because of its unerring ability to identify a faulted phase. Irrespective of the circumstances, for a nonfault condition, current into one phase of an unfaulted transmission line must equal current out the other end, except for the influence of distributed capacitance or shunt reactors. Faults involving multiple transmission lines pose serious interpretation problems for voltage-dependent systems because a fault on one circuit produces a voltage depression on all others. An "A–phase-to-ground" fault on one circuit, for example and a "B–phase-to-ground" fault on another may be interpreted as an A-to-B–phase fault by distance relays on both circuits. This devastating incorrect interpretation of fault type is not experienced by the segregated-phase comparison system.

A single-pole trip system utilizing this concept would correctly trip A-phase only and B-phase only, on the appropriate circuits, leaving four phases intact. Other systems would trip all three phases of both circuits.

No relaying is perfect, even setting aside the consideration of cost. Phase-comparison systems provide a purely pilot function; thus, they have no backup provision. This is not to say that a separate distance function cannot be included in a phase-comparison system, particularly considering the massive computational capabilities of present-day microprocessor-based relays. One version of the segregated-phase–comparison system activates a two-zone distance backup function whenever the channel deteriorates to an unacceptable performance level, where the pilot system is inoperative.

With phase comparison being entirely current-responsive, all current effects are important and must be interpreted correctly or nullified. Distance relaying on the other hand has voltage, which acts as a restraint, thereby lowering the significance of current aberrations (in exchange for accepting the influence of voltage aberrations). The energization of a transmission line produces a current transient. Although of substantial magnitude, it is also of high frequency, providing a means of distinguishing between that and the energization of a fault. External faults produce a discharge of this distributed capacitance of a protected line producing the appearance of an internal fault, but also being of high frequency and, thereby, being recognizable.

The steady-state, rated–frequency-charging current associated with a long EHV (Extra High Voltage) line is not to be ignored. It may be appreciable, and it has the appearance of an internal fault. Some provision must be included in the phase-comparison system to accommodate this effect.

Any "through" current is ignored by a phase-comparison system, but this may prove to be a detrimental characteristic when detecting high-impedance ground faults. Load current appears as an external fault and may mask a low-magnitude internal fault.

Inequitable distribution factors (different per unit zero-sequence and positive-sequence current flow for an internal fault) may cause improper sensing with a composite filter, phase-comparison system. With the filter output voltage being $C_1 I_{A1} + C_2 I_{A2} + C_0 I_{A0}$, it is influenced in part by load current (I_{A1}) and certainly by fault type (I_{A1} in-phase with I_{A2} and I_{A0} for an A-phase-to-ground fault and 180°, perhaps, between I_{A1} and I_{A2}/I_{A0} for a BC-ground fault). This leads to many different comparisons with the worst being imposed with a phase-to-phase–to-ground, predominantly positive-sequence contribution from one terminal, and predominantly zero-sequence contribution from the other. With an unwise choice of weighting factors, this leads to failure to recognize an internal fault. No such problem exists with a segregated phase-comparison system.

4. CURRENT DIFFERENTIAL RELAYING

Current differential schemes extend the phase-comparison principle by including magnitude as well as phase relationship in the comparison of the quantities at the two transmission line terminals. Again, the composite symmetrical component filter concept is used in one variety of current differential relay in the interest of including the influence of all three-phase and ground currents into one single-phase quantity, while limiting the information that must be transmitted between terminals. See Chapter 2 for a complete description of pilot wire relaying.

After the remote information is transmitted and converted to magnitude and phase, it may be compared with the similarly generated local phasor in a manner much like that in a differential relay. One form develops an operating quantity, which is the magnitude of the phasor sum of the local and remote quantities, and a restraining quantity, which is the sum of the magnitudes of the local and remote quantities. Tripping is initiated if $OP > 0.7$ REST.

Although useable for protecting a transmission line of any length, this system is particularly useful for protecting short lines, as well as circuits having a significant zero-sequence mutual impedance with other circuits. See Section 6.1 (Zero-Sequence Mutual).

5. DIRECTIONAL-COMPARISON RELAYING

The directional–comparison type of relaying uses the fundamental concept that directional units at all terminals of the protected line must agree that a fault is internal for tripping of circuit breakers to be initiated. Simple directional units are used (more frequently for ground) but phase directionality is virtually always determined by distance units. When using distance units, advantage is taken of their distinctive cut-off in reach and their insensitivity to load, thus providing additional security, and allowing other uses to be made of these units, such as time–distance backup protection.

5.1 Blocking Scheme

This widely used system utilizes overreaching distance units that detect all faults on the protected line (within the limits of their sensitivity). The pilot channel, which is generally on–off power-line carrier, is used in a blocking mode. Reception of a carrier signal from either the local or the remote transmitter blocks tripping by the overreaching distance unit. A phase or ground fault detector keys the carrier on. The carrier is turned off and held off by the operation of the overreaching forward-looking distance unit. Faults that occur on the line cause the carrier to be turned off at both terminals. Faults that occur at some location other than on the protected line cause carrier to be transmitted from one terminal, thereby blocking undesired tripping at either terminal. The essence of the scheme is shown in Figure 4-3. It is applied in electromechanical, solid-state, and microprocessor technology.

In the blocking scheme, it is customary to use the same carrier frequency at all terminals. Carrier transmission is then a general broadcast to all terminals associated with the protected line to refrain from tripping. For internal faults, 21P or 21NP (or 67NP) and I_0 operate at all terminals, and there is no carrier transmission.

5.1.1 Coordinating Timer

Figure 4-3 is simplified and shows only the elements pertinent to the fault X. The complete system consists of tripping and carrier-starting elements at all terminals. Many refinements have been incorporated in modern blocking systems. One obvious need is the coordinating timer shown in Figure 4-3b. This is necessary to ensure that an external fault is recognized, carrier is transmitted, and blocking is set up before the tripping relays can energize the trip coil. If the start relay is sufficiently fast (as it is, using the $\Delta I/\Delta V$ concept of modern microprocessor relays) compared with the tripping relay speed, no independent coordinating timer is required if a fast carrier system is used. With the trip and start devices having essentially the same operating speed, 4 ms is a representative coordinating time for this, but it is highly dependent on the nature of the channel in use. This time, necessarily, delays pilot tripping by a slight amount.

5.1.2 Backup

Many systems in service also utilize the 21P/21NP pilot function to drive a zone-2 timer. This may save some apparatus, but it has two disadvantages. First the reach of the pilot relays is restricted because of the requirement that a zone-2 relay cannot be allowed to overreach any adjacent line zone-1 relay and, second, there is a loss in the desired independence of primary and backup relaying. The preferred arrangement keeps primary and backup completely independent of one another from a setting and common component failure standpoint.

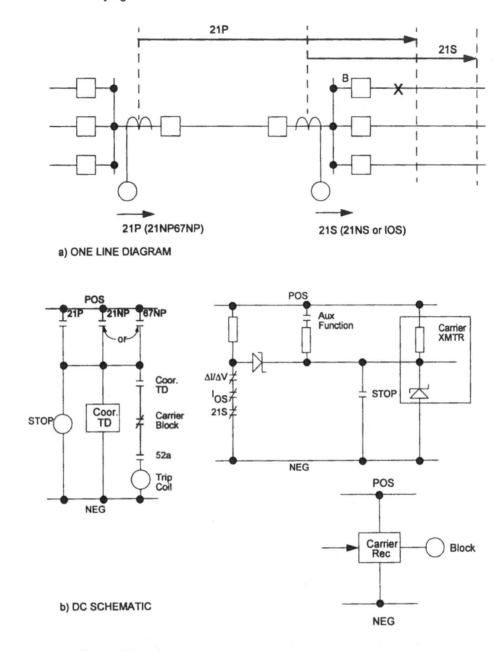

Figure 4-3 General concept of directional-comparison blocking scheme.

5.1.3 Channel Failure

Channel failure is not immediately recognized owing to its normal quiescent state. Unrecognized channel failure can result in overtripping because of the inability to transmit or receive the blocking signal for an external fault. Many systems are equipped with a carrier check-back scheme that periodically examines the integrity of the channel. Failure of the channel check produces an alarm, permitting the problem to be recognized and corrected before the need for the channel arises.

5.1.4 Traveling Wave Relays

Another form of directional comparison relaying that is able to provide extremely fast tripping uses the "wave-discriminator" principle. It compares ΔV, the change in voltage at the relay location with ΔI, the change in relay

c

current. The trajectory on the $\Delta V/\Delta I$ plot identifies the direction to a fault. Comparison, by way of a communication channel, of perceived direction from each transmission line terminal to a fault distinguishes between those that are internal and those that are external. This distinction can be made in a few milliseconds with the tripping time being

significantly influenced by channel speed. An "independent mode" trip may also be possible for long-line applications, eliminating the channel delay associated with the described "dependent mode."

Figure 4-4 displays the concept of the wave discriminator relay. Forward faults produce ΔV and ΔI (change in

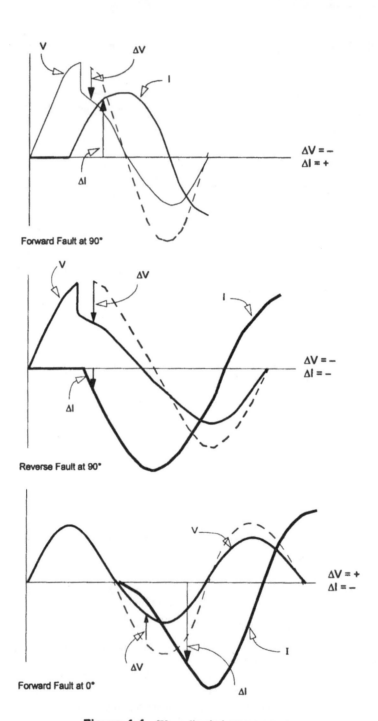

Figure 4-4 Wave discriminator concept.

sampled fault from one cycle previously) which are of opposite sign. Reverse faults produce ΔV and ΔI that are of the same sign. From this basic interpretation and through the use of a communications channel, very high-speed directional-comparison relaying is possible.

5.2 Transfer–Trip-Relaying

This relaying system uses a frequency-shift channel to allow continuous supervision of the channel. The two frequencies are called "guard" and "trip," with guard being the normally transmitted frequency. Two variations of this system are used: overreaching and underreaching. Which type is in use is identified by which setting is used on the relaying unit that keys the channel equipment to the trip frequency.

5.2.1 Permissive-Overreaching Transfer–Trip

This system, known as POTT, depends on overreaching phase and ground distance units to recognize the presence of a fault on the protected circuit and to initiate the transmission of a trip request to the remote terminal. Receipt of the trip frequency from a remote terminal and operation of the local overreaching phase or ground distance unit produces tripping (Fig. 4-5). Failure of the channel or excessive noise produces immediate blocking of tripping. The use of a tone system over a communications path other than the transmission line itself, or the use of adequate trapping for power line carrier, assures a high degree of probability of availability of the channel-for all fault cases, except for the condition of a bolted internal line-to-ground fault on the phase to which a power line carrier is coupled.

No "start" relays are required for the communications channel keying because a blocking signal is continuously transmitted. A frequency-shift channel is used with the guard frequency utilized in a blocking mode. The POTT scheme obtains the directional comparison at the two terminals through the requirement that the trip frequency *must* be received for tripping to take place. Keying of the transmitter is done by the 21P or 21NP.

The term *permissive* means that some other operation is required of the relaying at the terminal, in addition to responding to the receipt of the trip frequency. The term is redundant in POTT because an overreaching relay does the keying to the trip frequency; therefore, it is mandatory that a local relay confirm that the fault is on the protected line and not external to it. The 21P/21NP combination, thus, performs the dual function of keying the transmitter to the trip frequency and supervising tripping on receiving the trip frequency from the remote terminal.

This scheme has the advantage of requiring no channel "starting" function. Tripping for external faults is blocked by the presence of the "guard" signal. Guard is always transmitted except when the channel is keyed to trip. The lack of confidence in the ability to transmit a low-wattage signal through a fault that may be on the phase to which the signal is coupled has deterred the use of this system

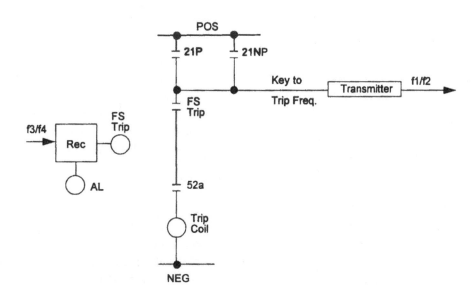

Figure 4-5 General concept of overreaching transfer–trip.

with power-line carrier. It is used, generally, with tones on a voice channel on a wire-line or over microwave. It has the further advantage of requiring no coordinating time delay. The blocking signal is always in the proper state for blocking trip for external faults.

Impulse noise causes the channel trip output to be blocked temporarily. Channel failure also causes blocking of all pilot tripping until the channel has been properly restored.

5.2.2 Underreaching Transfer–Trip

Two subcategories are used for this type of protective relaying system. The first requires no supervision of a received trip request from a remote terminal and tripping is immediate. Having been identified as an internal fault within the reach of the remote zone-1 distance unit, no further assurance is required for the need to trip the local breaker. This places heavy dependence on the security of the frequency-shift communications system. This system is called a *direct*-underreaching transfer–trip scheme (DUTT). The *permissive*-underreaching transfer–trip scheme (PUTT) has an additional layer of security through the use of the zone-2 (overreaching) units to supervise tripping when the trip frequency is received from the remote terminal.

Direct-Underreaching Transfer–Trip

The DUTT system is not truly a directional-comparison pilot-relaying system, but it is usually classified this way. With zone-1 phase and ground relays available at both transmission line terminals, most faults will be detected (60% of them using an 80%-reach setting and 80% using a 90%-reach setting for the zone-1 distance relays) by the zone-1 units at *both* terminals with no need for a received signal from the remote terminal (Fig. 4-6). The pilot channel contributes high-speed tripping for faults that occur in the end zones, those 10 or 20% regions on the transmission line that are immediately adjacent to the two stations. With the channel out of service, tripping will always be at high speed at one or both terminals, depending on fault location. For end-zone faults, dependence is placed on zone-2 time-delayed tripping at the terminal remote from the fault. This system can be modified (with an additional sacrifice) so that channel failure allows high-speed tripping by the zone-2 unit. Security is sacrificed at the expense of dependable high-speed tripping. Some overtripping will occur if this variation is used, and considerable thought should be given to the implications of overtripping as compared with possible delayed tripping at a terminal remote from the fault. Where an adequate high-speed backup relaying system exists, this variation (eliminating

zone-2 timing on recognition of channel failure) should not be used.

Permissive-Underreaching Transfer–Trip

As the term "permissive" connotes, some permission is required by a local terminal for tripping to take place in response to a received trip signal over the communications medium. This permission is usually granted by a set of overreaching (zone-2) phase and ground distance units. High-speed tripping for end-zone faults, then, requires that one end trip immediately through the operation of a zone-1 relay unit, and that the other end trip on the simultaneous operation of a zone-2 unit with the receipt of a trip frequency from the far end.

This system is classified as an underreaching system, even though some tripping is dependent on the operation of a zone-2 unit, because the channel is keyed to the trip frequency by the underreaching (zone-1) units.

5.2.3 Unblocking Relaying

An unblocking transmission line relaying system is fundamentally a permissive (all overreaching systems are permissive) overreaching transfer–trip scheme that incorporates additional logic that allows tripping for the period immediately following loss of channel, if distance units identify the location of a fault to be within their reach and in the forward direction. This allows the use of the POTT principle using power line carrier without the shortcoming associated with an internal fault that produces channel failure. The use of the unblocking scheme presumes that, with proper trapping, no channel failure will occur simultaneous with any fault except an internal fault. The system operates completely as a POTT scheme at all times when the channel is operative.

When a fault occurs and the channel is lost as a result of it, a period, such as 150 ms, is allowed following the loss of channel in which tripping can occur as a result of zone-2 distance unit operation. Following this time, all pilot tripping is blocked. Channel failure for this or any other condition is acknowledged by sounding the alarm.

5.3 Other Systems

Several relaying systems have been used to approach the performance of a pilot system without actually using a communications channel. Their serious shortcomings are expected and accepted in the interests of economy and at the sacrifice of performance. Two such systems are

1. Load-loss trip
2. Zone-1 extension

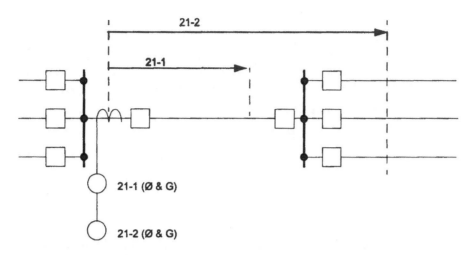

ONE LINE DIAGRAM

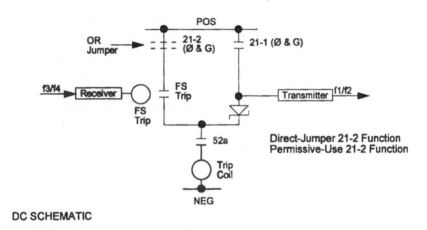

DC SCHEMATIC

Figure 4-6 General concept of underreaching transfer–trip.

5.3.1 Load-Loss Trip

The load-loss trip system takes advantage of the fact that clearing of all types of end-zone faults by a zone-1 relay causes the unfaulted phase currents to drop to essentially zero at the terminal away from the end zone. Detection of the presence of prefault load and operation of a zone-2 relay along with detection of loss-of-load in one or more phases produces high-speed trip. Aside from the qualifiers, the overall results are the same as those that would occur with a PUTT scheme. For faults occurring in the overlapped region for the zone-1 relays, tripping occurs independently of this load-loss logic.

For load-loss trip to be activated, prefault load must exist, load taps must be minor, and the fault must not be three phase. Other provision must be included for three-phase fault clearing.

5.3.2 Zone-1 Extension

The zone-1 extension scheme is incorrectly named, but historical precedent encourages the continued usage. More correctly this should be called a "zone-2 contraction" scheme. Faults occurring within the reach of zone-2 (phase or ground) produce *immediate* tripping of the breakers at both terminals of the protected line. Those faults that are internal to the protected line are cleared immediately without the need of a pilot channel. Being an overreaching scheme, some forward external faults will be seen by the zone-2 relays. Without the aid of a pilot channel to provide

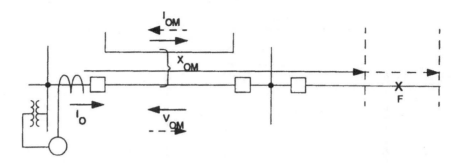

Figure 4-7 Pilot distance reach variation caused by zero-sequence mutual.

a blocking function, overtripping will occur. A reclosing function is mandatory with this scheme. When tripping occurs by the relays on the faulted line as well as by the relays on an adjacent line, both relays then have their reach contracted to a zone-1 setting. When reclosing takes place, only the relays on the faulted circuit operate or, if the fault were temporary, all relays reset to their normal zone-2 reach.

6. PROBLEM AREAS

Many power system phenomenon produce detrimental effects on some relaying systems, but not on others. This section describes some of these, and points out how they may influence the choice of one pilot system over others.

6.1 Zero-Sequence Mutual

Induced zero-sequence voltage in one transmission line, resulting from zero-sequence current in another (and the mu-

tual impedance between the two) seldom has any beneficial effects. Figure 4-7 shows that the induced voltage may be in either direction, depending on the relative directions of current in the protected line and in the adjacent line.

6.1.1 Effect on Current-Differential Relaying

Current magnitude is influenced by mutual impedance, but internal faults appear to be internal, and external faults appear to be external, irrespective of the extent of a mutually induced voltage. A ground fault on a adjacent circuit will cause zero-sequence current to flow. All of the attributes of external faults are present and proper restraining takes place.

6.1.2 Effect on Phase-Comparison Relaying

Phase-comparison relaying is not affected by zero-sequence mutual impedance for the same reasons as those for current differential. With zero-sequence mutual impedance, internal faults appear to be internal and external faults appear to be external—always.

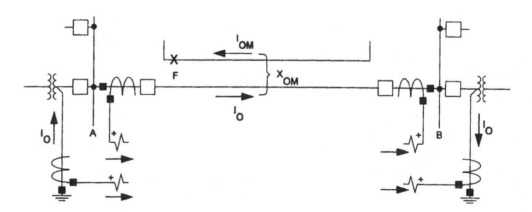

Figure 4-8 Directional comparison ground–false trip caused by zero-sequence mutual.

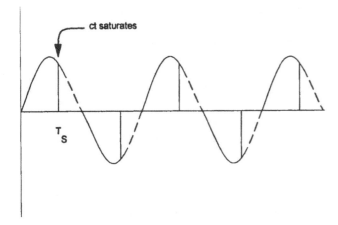

Figure 4-9   The ct output during symmetrical saturation.

6.1.3   Effect on Pilot-Distance Relaying

As Figure 4-7 portrays, zero sequence mutual impedance can cause ground distance relays to underreach or overreach, depending on the direction of the zero-sequence current in the adjacent circuit.

Figure 4-8 shows a much more devastating phenomenon that results from a ground fault on an adjacent transmission line. The zero-sequence voltage induced in the protected line A–B by zero-sequence current in the adjacent line causes zero-sequence current to flow through the line. This causes zero-sequence current (or voltage)-polarized directional relays protecting line A–B to operate. Unfortunately they operate at both transmission line terminals and for as long as the fault persists. False tripping results. Negative-sequence directional units are immune to this.

6.2   Current Transformer Saturation

Adequate current transformers (ct's) are vital in all pilot applications because security is needed for all external faults. To design a relaying system to trip at high speed for all internal faults is a comparatively simple task. To equip it with the ability to refrain from tripping for all circumstances, other than an internal fault, poses a real challenge.

Current transformers experience errors in magnitude and phase angle. The most serious errors are those attributable to severe saturation of the core of the ct. It is felt that a review of this phenomenon as it relates to the behavior of modern solid-state and microprocessor relays would be useful.

The simplest means by which the effects of saturation can be demonstrated is through the use of Figure 4-9. This represents the waveform of current that would be delivered to a relay while experiencing saturation at time T_S and on each subsequent half-cycle. No effect of dc offset in the primary current is considered in this figure. After T_S and until the next zero crossing, zero current is assumed to be delivered by the current transformer to the relay. This is referred to as "symmetrical saturation." The process repeats each half-cycle.

It is pertinent to examine the fundamental (60-Hz) content of this distorted waveform. With use of Fourier analysis, the curve of Figure 4-10 was developed. It shows that symmetrical saturation occurring substantially after the peak does not produce extreme reduction of the fundamental (60-Hz) component.

Another extremely important influence in protective relaying is the current transformer asymmetrical saturation that may occur as a result of the dc component in the fault

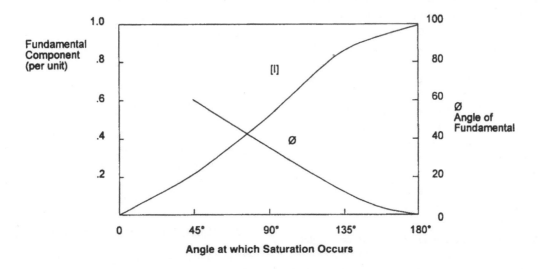

Figure 4-10   Fundamental output of ct during symmetrical saturation.

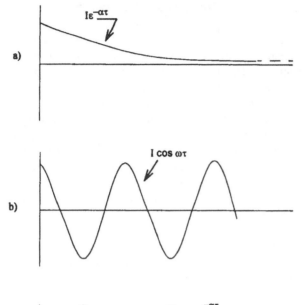

Figure 4-11 Fully offset current with decaying dc.

current. For internal faults, the system should be designed in such a way that tripping is sealed before ct saturation occurs. This is related to V_K/IRT with V_K being the ct knee voltage, I the maximum symmetrical secondary current in amperes rms, R the total secondary loop resistance in ohms, and T the dc time constant of the primary in cycles (60 L/R). [Protective Relaying, Theory and Applications (ABB/Marcel Dekker), Chap. 5, Fig. 5-11, describes this in detail.]

For external faults, assurance must be obtained that similar behavior will occur by the ct's at the two line terminals (good or bad) or that the nature of the relaying system will accommodate the maximum differences that can occur. For an external fault the ct's encounter essentially the same current, the same dc time constant and, having the same ratio, will behave similarly if their burdens and remanent flux are similar.

The waveform of Figure 4-11c was examined to determine the influence that a Fourier filter would have on sensing by a protective relay. No saturation is present, and the waveform is considered to have been delivered to the mea-

suring device in its entirety. By using eight samples per cycle, a fundamental value of current can be calculated. The decaying exponential will contain sine and cosine elements that will cause the measurement of the fundamental (60-Hz) component to have error in magnitude as well as angle. The extent of these errors is dependent on T, the dc time constant, and is shown in Figure 4-12. To minimize these errors, the first two samples are discarded before the Fourier products are produced.

6.2.1 Effect on Current Differential Scheme

This scheme is capable of misoperation at the occurrence of ct saturation, but it is not likely to do so. For external faults, with restraint being related (in the LCB-II) to the sum of the magnitudes of the local and remote filter output voltage, this restraint will be substantial. The operating quantity is the magnitude of the phasor sum. If we now ignore all but ct errors, and consider that one set transform without error and the ct's at the other end of the transmission line deliver a deficient current magnitude to the relays at that point, the behavior of the system may be examined by investigating the operating and restraint quantities:

$$V_{OP} = |V_{F1} + V_{F2}| \qquad (4\text{-}1)$$
$$V_R = 0.7 \, [|V_{F1}| + |V_{F2}|] \qquad (4\text{-}2)$$

If the deficient current for this external fault is 180° out-of-phase with the perfectly transformed current I, at the other terminal and has a magnitude KI,

$$V_{OP} = |V_{F1} - KV_{F1}| = V_{F1} (1 - K)| \qquad (4\text{-}3)$$

and

$$V_R = 0.7[V_{F1} (1 + K)]$$

With some margin, the relay will reliably restrain under this consideration if:

$$0.7 (1 + K) > (1 - K) \qquad \text{or} \qquad K > 0.16$$

If the deficient current is now assumed to be at 90° relative to the perfectly transformed current, restraint is assured if $K > 0.246$.

Thus, it can be seen that current at the location where the deficient ct's exist may be as low as 25% of the current at the other terminal and still not misoperate for an external fault. The current differential relay is very secure in the presence of inequitable saturation.

6.2.2 Effect on Phase-Comparison Relaying

Phase-comparison relaying is influenced by ct saturation in many ways, depending on the nature of the scheme. One scheme (MSPC) will trip reliably at high speed for an in-

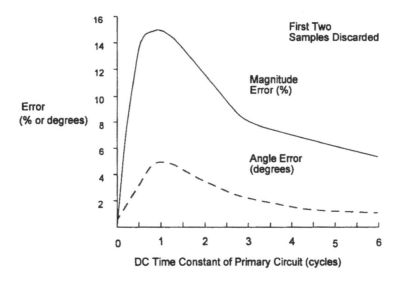

Figure 4-12 Influence on fundamental of maximum dc offset using Fourier filter.

ternal fault if *either* terminal is able to trip before saturation occurs, owing to the dc component in the fault current. It then clamps to the trip-positive key, allowing the other terminal to trip for any positive (or zero) value current.

An internal fault with severe saturation of the ct's at one terminal is shown in Figure 4-13. For the example shown, tripping occurs with considerable margin.

Even though saturation might occur, in a worse case, very early in the cycle, tripping of all terminals would still occur at high speed, provided that ct's at one location deliver current to the relay with reasonable fidelity long enough to satisfy a 3-ms timer (4 ms for ground).

Figure 4-14 shows an example in which an external fault exists, offset keying is used (MSPC), and saturation of the ct occurs at one terminal only, because of the dc component of the fault current. Even with this severe difference in behavior of the ct's at the two terminals, the system is secure, although "holes" in the current permit undesired timing to take place.

Note that symmetrical saturation (see Fig. 4-9) may produce little effect on relays (such as the MSPC) which use a 240-Hz cutoff antialiasing filter because only the fundamental is presented to the sensing elements. Figure 4-10 describes the influence on this fundamental.

6.2.3 Effect on Distance Pilot Schemes

Current transformer deficiencies, in general, cause distance relays to see a lower effective current than they would see otherwise. This causes them to reach a shorter distance than they would if there were no ct saturation.

Relays set to reach all faults on a protected transmission line will, in response to this effect, not fail to operate, but rather, will sense that the fault is farther away than it actually is. This will cause an electromechanical pilot distance scheme with minor ct saturation to trip perhaps as

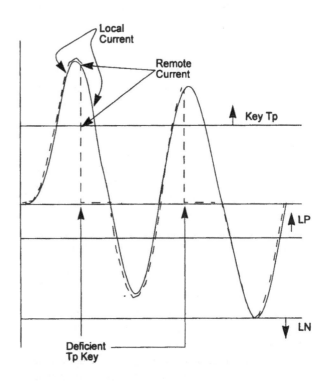

Figure 4-13 Effect of different ct saturation on offset-keying phase-comparison (internal fault).

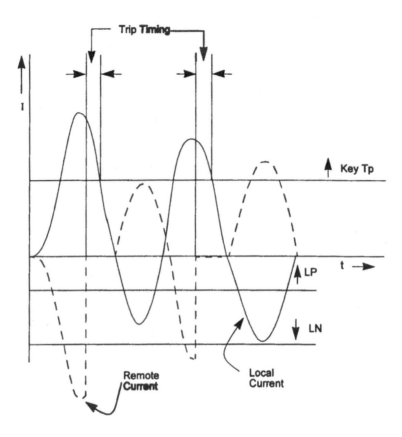

Figure 4-14   Effect of different ct saturation on offset-keying phase-comparison (external fault).

much as 10–20 ms slower because of its dependency on energy level at the relay. A numerical relay encountering the same conditions will experience the same "reach-shortening" effect, but will not be subject to the operating time variations.

In a "directional comparison blocking" scheme, it is imperative that a "channel start" function take place at the terminal nearest the fault for all external faults for which the overreaching tripping relays at the terminal farthest from the fault are able to respond. Differences in ct performance at the two terminals can prevent the near "channel start" relay from operating for a fault for which the remote pilot trip relay operates. If this should happen a false trip will occur. This is attributable to the two different RMS current magnitudes delivered to the relays at the two locations. Directional overcurrent relays (tripping and carrier start) are more susceptible to this problem than are distance relays, although both are vulnerable.

Microprocessor distance relays generally use a Fourier filter in isolating the fundamental component of current for distance measurement. Figure 4-12 depicts the influence of this treatment on the apparent fundamental (60-Hz) com-

ponent of a current waveform, such as that of Figure 4-11c. Precautions must be taken with zone-1 relays to assure that the influence on the fundamental component by the dc component is taken into account. The possible effect of dc on the apparent fundamental component is not nearly so extreme as its effect on the RMS value, to which other types of devices respond.

6.3   Coupling Capacitor Voltage Transformer Transients

Coupling Capacitor Voltage Transformers (CCVTs) are excellent, economical substitutes for iron core voltage transformers. They do, however, introduce transients of their own, causing relays to receive voltage that is not truly indicative of that on the power system. In general, the higher the capacitance of the device, the better the fidelity of the transformer.

6.3.1   Effect on Current Differential Scheme

Current differential schemes use no voltage supply; therefore, they are not influenced by this phenomenon.

6.3.2 Effect on Phase Comparison

Basic phase comparison relaying uses no voltage input, although some variations of this scheme have in the past required its availability. "Distance" phase comparison and "combined" phase comparison used distance relays, but they are not considered here. Modern phase comparison relays (such as MSPC) contain a backup, stepped distance system that is activated by the loss of channel. For time-delayed trips the CCVT transient is not significant.

6.3.3 Effect on Distance Pilot

"Blocking," "unblocking," and "overreaching" schemes use, for their basic-sensing function, overreaching, distance relays. The extent of the overreach is not critical. With the overreach set to assure that the worst CCVT transient still allows complete line coverage, there is no detrimental effect on this type of system. This may require a much longer reach setting on the overreaching relays than would otherwise be necessary. Because the zone-2 time trip function has a limited reach possibility, zone-2 and pilot trip distance functions should be independently settable.

Underreaching transfer–trip-relaying systems must have their zone-1 distance settings chosen in such a way as to avoid overreaching the next bus during the worst-case CCVT transient. If this effect is too extreme, the zone-1 relays at the two terminals will fail to overlap, causing a hole to exist in the pilot scheme.

6.4 Channel Problems

Relaying channels are quite diverse and experience noise, deterioration, and failure. Any pilot system must be able to circumvent these problems by trip-blocking, error correction, or reverting to backup that is not channel-dependent. Power-line carrier uses the protected transmission line as part of its communications path. Faults on the line may not, but must be expected to, cause loss of channel. On–off carrier-relaying systems use the off state as a trip condition (accompanied by other logic). There is no normal transmission of carrier and no means of continuous channel monitoring.

Frequency-shift carrier continually transmits a guard signal that can be supervised. Distance relays shift to a trip state. The inadequacy of the trip state may not be detected during guard transmission. Phase-comparison relays continually shift from "mark" to "space." Faults may cause loss-of-channel.

Microwave, fiber-optic, and pilot wire channels do not use the protected line as part of the communications path, but are subject to their own variety of reasons for failure.

6.4.1 Effect on Current Differential Relaying

Several options may be exercised when channel trouble is detected. The system may be allowed to revert to a simple overcurrent system, with full advantage still taken of the symmetrical component filter, providing sensitive ground detection and suitably desensitized positive sequence detection. If this is not desired, the system can reliably, automatically be removed from service until the channel is restored. This places total dependence during the channel outage on backup devices.

6.4.2 Effect on Phase-Comparison Relaying

The basic phase-comparison function is incapacitated when the channel is lost. Modern microprocessor phase-comparison relays revert to a 2-zone–step distance backup when channel trouble is detected.

Power-line carrier (on–off) phase-comparison trips if a fault is detected and carrier becomes lost. Frequency-shift phase comparison operates with proper phase coincidence or for 150 ms following loss of channel (unblock logic), then followed by lockout of the relaying system. These systems are vulnerable to a false trip when external fault detection and loss of channel occur simultaneously. Adequate line traps and adequately spaced drain coil and tuner gaps prevents this.

6.4.3 Effect on Distance Pilot Relaying

Blocking schemes must receive a carrier (or other) signal for external faults. POTT schemes must receive a signal from the other terminal for internal and external faults. Unblocking schemes must receive a signal from the other terminal for external faults.

Underreaching transfer–trip schemes do not require the channel for any external fault security, nor for faults occurring in the usually substantial zone of overlap of the zone-1 relays. The channel is required only for end-zone faults to assure high-speed tripping of both terminals.

Loss of channel is not detected immediately in a carrier blocking scheme. Therefore, it is more susceptible to a false trip. POTT carrier schemes are susceptible to failure to operate in the event of channel problems.

6.5 Power System Swings

Swings occur following faults and switching. Changing voltage and current are applied to the relays. Stable swings are self-correcting and *must* be ignored. Unstable swings require action at some locations and prevention of action at others. To do this on a discriminating basis is the goal.

6.5.1 Effect of Current Differential Relaying

Swings appear as a "through" phenomenon and are ignored by these relays.

6.5.2 Effect of Phase-Comparison Relaying

These relays also acknowledge swings to be an "external" phenomenon and restrain properly.

6.5.3 Effect on Distance Pilot Relaying

Distance relays may respond to the reduced voltage and increase current that they encounter during a swing. The ohmic trajectory that is manifested during a swing may be similar to S_1 or S_2 in Figure 4-15. Zone-1 is obviously less susceptible to operation on swings than the pilot distance relay $21P_A$, simply because of its smaller characteristic circle. Thus, the PUTT system is more secure than the POTT scheme against swing trips.

Swing S_2 enters $21P_A$ circle before it enters $21S_B$. For the blocking scheme, a false trip results. In the unblocking or POTT scheme, continuous transmission of a blocking signal from station B to station A is uninterrupted by this. These two schemes are, then, more secure than the blocking scheme and, incidentally, do it with less equipment (or fewer algorithms).

6.6 Three Terminal Lines

Three terminal lines inject the problems of possible outfeed for internal faults, infeed effects, possible weakfeed conditions, and channel complications. See Chapter 5 for a detailed description of three-terminal line relaying.

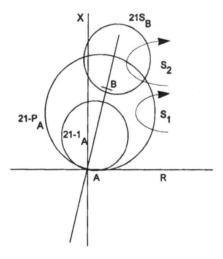

Figure 4-15 Influence of swings on distance pilot relaying.

6.6.1 Effect on Current Differential Relaying

The principal problem relates to the channel. Minimal contribution from one terminal to an internal fault or even small outfeed causes no problem with this system. The V_F voltage from each terminal must be transmitted to all terminals.

One terminal capable of delivering a large contribution to an internal fault is able to produce high-speed simultaneous tripping at *all* terminals.

6.6.2 Effect on Phase-Comparison Relaying

Because of the difficulty of comparing current phase relationship at one location with the phase angle of the phasor sum of the currents at the other locations, this scheme is limited in its application.

With no outfeed possibilities, the on–off carrier scheme is usable, with all terminals tripping for internal faults during the off period. With frequency-shift systems, each transmitter must communicate continuously with both remote receivers, somewhat complicating the channel requirements.

With outfeed at one terminal for an internal fault, this system is unsuitable.

6.6.3 Effect on Distance Pilot Relaying

The communications requirements are simplest in the carrier-blocking schemes. The nature of the system is that the absence of carrier allows distance relays to trip. All of the carrier systems, then, operate at the same frequency. Only one receiver is required at each terminal. Any terminal requesting block gets it at all terminals. Any terminal having no current cannot block tripping at the other terminals. If the maximum outfeed for an internal fault is smaller than the starting current level, the blocking scheme will trip correctly for this case.

POTT and unblocking schemes use frequency-shift channels. Because the trip frequency must be received (or the channel lost in the unblocking system) when an internal fault occurs, all terminals must independently recognize the fault and send a trip request (shift to the trip frequency) to all terminals. Separate receivers for each remote terminal are required. Thus, these systems have more complicated communications channels for three-terminal applications and are dependent on all terminals being able to recognize internal faults.

PUTT systems bow to DUTT (direct underreaching transfer trip) systems when there is a weak source terminal. Any zone-1 relay operation keys to the trip frequency, and without any distance relay operation at the two remote stations, immediate high-speed tripping takes place at all locations.

This system is sensitive to the location of the tap for the third terminal. No zone-1 relay can be allowed to overreach any bus. A transformer at the tap often eases this problem.

6.7 Evolving Faults

Faults having one character initially and changing to a different state subsequently are difficult or impossible for relaying systems to handle. These faults may, after a short time, involve other phases or even other transmission lines. They initially may be external and become internal.

6.7.1 Effect on Current Differential Relaying

"Current differential" systems derive their operating quantity from the phasor sum of the symmetrical component filter outputs at the two terminals. Any combination of fault types involving a net internal fault contribution will produce tripping. However, the use of the composite filter may produce difficulties with some combinations of faults. An external ground fault simultaneous with an internal phase fault may produce undesired strong restraint to accompany the operating quantity. Zero-sequence is usually heavily weighted compared with the positive- and negative-sequence weighting and may block tripping for the internal fault.

6.7.2 Effect on Phase-Comparison Relaying

The "segregated phase-comparison" system is without equal in this category. In an untapped transmission line, any current that is present in one phase at one terminal that is not present in the same phase at the other terminal is internal fault current or changing current, and there is no difficulty in distinguishing between the two. Any combination of internal faulted phases will be recognized as an internal fault, irrespective of the involvement of the same or other phases in an external fault and irrespective of the presence of an internal or external series capacitor with or without flashed gaps.

6.7.3 Effect on Distance Pilot Relaying

Faults, evolving external to internal can be troublesome, possibly causing delayed recognition or even failure to trip. POTT schemes use a transient-blocking scheme to make certain equitable resetting takes place during the current reversal that may occur owing to sequential clearing of external faults. This requires the introduction of a trip time delay for internal faults following the establishment of transient blocking.

Simultaneous internal and external faults on different phases may produce inappropriate currents and voltages for proper impedance measurement, and the distance relays fail to trip.

6.8 Stub Bus Faults

With a line disconnect open in a two-breaker scheme (ring or breaker and a half), such as shown in Figure 4-16, faults on the bus section between the breakers may be difficult to detect with a pilot-relaying system. The line disconnect separates the two system segments, but the channel may

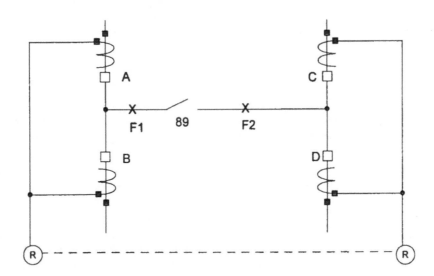

Figure 4-16 Pilot relaying with 89 open.

still be intact. Faults occurring on one side of the disconnect will not be recognized by the relays on the other side. The necessary cooperation between terminals is lost. The qualification that an 89b auxiliary switch (closed when the disconnect is open) be available makes several relaying systems suitable, whereas, otherwise their performance would be deficient.

6.8.1 Effect on Current Differential Relaying

With all breakers closed and 89 open, a fault at F1 will be recognized because of the generation of equal operating quantities at each relay. Breakers A and B are tripped to clear the fault. However, breakers C and D are also tripped, even though no fault current is contributed through them. This undesirably splits the system. Options are available in this system (LCBII or REL356) to prevent this and yet allow detection and clearing of faults F1 and F2 without overtripping. It requires an input to the relaying system of a "b" switch on device 89 to identify its opening. This "fails" the channel, to allow the relay at the far terminal to revert to overcurrent-only.

6.8.2 Effect on Phase-Comparison Relaying

Microprocessor phase-comparison relaying accepts the 89b input and generates and transmits to the remote station a code to identify the open condition. With 89 closed, normal relaying takes place and faults F1 and F2 cause all four breakers to trip. With 89 open, F1 produces tripping of A and B only, whereas F2 produces tripping of C and D only.

6.8.3 Effect of Distance Pilot Relaying

For a blocking scheme, discriminating tripping takes place even though 89 is open. F1 is cleared by opening A and B. F2 is cleared by opening C and D. Carrier transmission may take place for either fault owing to the sensing of ΔV, but it will last only 32 ms. For POTT and unblocking protective schemes, no tripping takes place with 89 open because, for example, F1 does not operate the overreaching distance relay at CD; therefore, transmission of the blocking (guard or space) signal from CD continues. This problem can be overcome by using the 89b contact. Received trip can be simulated locally when the 89b contact is closed to take care of fault F1, Keying to trip frequency by the 89b contact can arm the remote pilot system to be able to trip by distance relay operation for fault F2.

PUTT systems perform satisfactorily for all cases except a fault near 89 and outside of the zone-1 reach from CD. Fault F1 is cleared by zone-1 and the channel is keyed to trip. The permissive system requires that zone-2 at CD *permit* tripping. As it does not detect fault F1, C and D

properly remain closed. A *direct* underreaching protective system would overtrip for this same case.

6.9 Large Load and Limited Fault Current

For load to flow, there must be an angle between the positive sequence "sources" at the two ends of a transmission line. When an internal fault occurs, the contributions from the two ends of the line are out-of-phase. The extent of the influence is dependent on the magnitude of the load flow and the length of the line.

6.9.1 Effect on Current Differential Relaying

Normally load is a "through" positive-sequence phenomenon. Sensitive detection of a high-impedance ground fault by weighting of the zero current effect may be masked by very large load flow, which continues during the internal fault. Positive-sequence can be removed from the filter output, but this then places dependence for three-phase fault recognition on other devices. In general, the level of fault resistance that must be accommodated compared with the maximum value of load current is such that this need not be performed.

6.9.2 Effect on Phase-Comparison Relaying

A similar problem to that which exists for current differential is also present in phase-comparison relaying. Through-load current appears as outfeed. Essentially, for an internal fault, one end sees this fault contribution plus load current flowing into the line, and the other sees its fault contribution minus load current flowing into the line. One form of segregated-phase comparison relaying will respond correctly with 5 A of through-load current and 1 A of internal fault current. It also has a $3I_0$ subsystem that is not masked by load.

6.9.3 Effect on Pilot Distance Relaying

Load does not influence the reach of a distance relay at its maximum sensitivity angle. However, it will influence the polarizing voltage and, with fault resistance, a substantial change in reach may occur. Overreaching occurs, in general, with load flow in the trip direction, and underreaching occurs with load flow in the nontrip direction.

Very large load flow can get into the trip characteristic of distance relays that are responsive to positive-sequence quantities and that are applied in long-line applications. This is screened out as a standard course in microprocessor relays (such as REL302) by utilizing a blinder relay to block tripping for very large high-power factor loads.

Table 4-1 Pilot System Comparison

Performance in the presence of	Pilot system					
	Current differential	Phase comparison	POTT or unblock	Blocking	PUTT	DUTT
Mutual	SUP	SUP	RSS	RSS	RSS	RSS
ct saturation	SAT	SAT (Q)	SAT	SAT (Q)	SAT	SAT
CCVT transients	SUP	SUP	SAT	SAT	RSS	RSS
Channel problems	SAT (Q)	SAT (Q) (with inherent backup) U (composite filter)	POTT-U UB-SAT (Q)	SAT (Q)	U	U
Swings	SUP	SUP	RSS	RSS	RSS	RSS
Three-terminal, no outfeed	SUP	SAT (Q)	SAT (Q)	SUP	RSS	RSS
Three-terminal, with outfeed	RSS	U	U	RSS	U	RSS
Evolving faults	RSS	RSS (composite filter) SUP (SEG φ)	RSS	RSS	RSS	RSS
Stub bus	SAT (Q)	SAT (Q)	SAT (Q)	SAT	U	U
Load and high RG	RSS	RSS	SAT	SAT	U	U

SUP, superior; SAT, satisfactory; SAT (Q), satisfactory but qualified; RSS, requires special study; U, unsatisfactory.

Overreaching applications (POTT, unblock, blocking) are not reach-sensitive, and the load effect is not detrimental. Use of directional overcurrent for the ground pilot function provides very sensitive ground fault detection and avoids the influence of load. PUTT systems are highly susceptible to misoperation or excessive shortening of coverage owing to this effect.

In the presence of large load current, high-resistance ground faults are best detected by sensitive zero- or negative-sequence overcurrent relays. Underreaching pilot systems require the use of a distance relay (or element, or algorithm) for the ground function. Sensitive ground detection is not compatible with zone-1 distance ground relaying.

7. TABLE 4-1

Table 4-1 summarizes the influence of certain power system phenomenon on the most prominent pilot relaying systems. Some of the classifications are arguable, but those that are described as superior in the presence of certain transient conditions truly do have distinct advantage over other relaying systems, and those that are described as unsatisfactory do have serious shortcomings in the area described.

Among these systems that are described, none leaps out as being the overwhelming choice for all power system configurations, but the table should be of some help when certain difficulties are foreseen and relative behavior is being examined. The list of trouble sources is certainly not exhaustive, but the prevalent influences are covered.

5

Three-Terminal Line Protection

WALTER A. ELMORE

1. INTRODUCTION

In theory, a transmission line is a circuit, intended to transmit bulk power only, with no taps. A subtransmission line is intended to supply power to multiple distribution stations. With the constant pressure toward more stringent economy, many transmission lines become tapped to serve major leads at some point along the route. Many subtransmission lines have sources at both ends, in the interests of better service continuity. Because of these two tendencies many important circuits are, and will be, multiterminal. From a relaying viewpoint, two-terminal lines are straightforward and may be protected without undue deliberation. However, three-terminal lines impose critical restrictions and may require compromises in the relaying, with a resulting performance that is less than satisfactory.

The one-line diagram of Figure 5-1 describes the configuration referred to as a three-terminal line. Lines having more than three terminals impose conditions on the relaying that cannot be satisfied unless certain unique characteristics exist.

2. TYPES OF TERMINALS

Various types of power system connections exist, some of which may either alleviate or make more difficult the re-laying problem. Some of the obvious things that are influential are the following:

1. The transformer connection
2. The transformer grounding
3. Generation at the tap
4. Load at the tap
5. Outfeed

2.1 The Transformer Connection

Most transformers in a tapped circuit have either a wye–delta or a delta–wye connection. In either case, the delta winding manifests a zero-sequence current block. However, such a transformer with the wye–neutral grounded is a zero-sequence "source" (actually the fault or other discontinuity is the source, and the transformer is a path for the flow of zero-sequence current), and the wye winding will contribute current to ground faults.

2.2 Transformer Grounding

For a ground fault on the high-voltage system, a high-voltage wye will continue to provide a path for *zero-sequence* current flow until the last *positive-sequence* source is disconnected.

With the delta on the high-voltage (HV) side, it will contribute no zero-sequence current, irrespective of

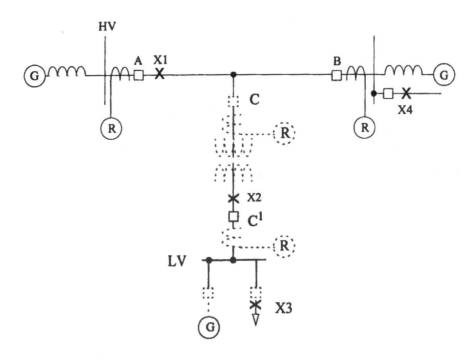

Figure 5-1 Three-terminal line.

whether or not the positive-sequence sources are connected. This difference must be acknowledged in choosing a relaying system. With generation at the tap, and a fault at X1 (see Fig. 5-1), the fault will continue to be energized after A and B open with major current flow if the transformer is wye–delta (i.e., HV wye) and minor (capacitive) current flow if it is delta–wye.

2.3   Generation at the Tap

For the fault at X1, with the delta–wye (i.e., HV delta) transformer, opening A and B completely de-energizes the fault if there is *no* generation at C. If there is generation at C, opening A and B leaves the fault energized with a good likelihood of overvoltages caused by restriking of the fault arc. It must be understood that, even though generation is present at C, there is some possibility that this generation will be shut down at the time a HV fault occurs. With multiple generators at the tap, the dependability of the tap as a positive sequence source is very high, but the strength of the source is variable.

Generation at the third terminal provides a useful contribution to a line fault, with which relaying can be accomplished. However, the nature of distance relaying is that it utilizes volts (V) and amperes (A) to form a measurement, and for this the contribution from the third terminal produces a voltage drop that influences the voltage at, for ex-

ample, location B in Figure 5-2. However, the current from C which produces this additional voltage drop is not accessible at station B; therefore, a proper interpretation of the distance to the fault is not possible. This is the well-known "infeed effect."

2.4   Load at the Tap

A load on the bus at the tap is significant only as it appears as an outfeed condition, with or without a fault on the high-voltage system. Care must be used in selecting the relaying scheme and in judging the influence of this load. If the load is small enough, compared with the level of current for a fault on the high-voltage system, it may be ignored. Industrial load is predominantly motors (some very large), and they will contribute to a fault as if they were generators for a period as long as their rotors spin and their fields are supported. This fact may assist in clearing internal line faults, but this is not generally considered to be a factor on which we can rely.

With motor loads, it is imperative that they not be re-energized once they have been separated from their power supply until the residual voltage has decayed to something like 25% of normal or less. Where reclosing is used, this means that high-speed separation (tripping) of the third terminal is required or reclosing must be delayed. When

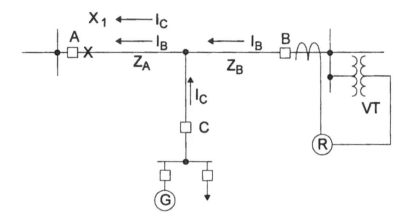

Figure 5-2 Infeed effect.

the third terminal is not equipped with line relaying, and reclosing must be at high speed, underfrequency or undervoltage relaying, or both, must be used to trip the supply breaker to protect the motors and their driven loads against sudden out-of-phase reenergization.

2.5 Outfeed

Certain system connections allow the current from the third terminal to flow away from the protected line for a fault on the line. Figure 5-3 shows the relative current directions for a fault near one line terminal. The current in Z_T may be as shown if the contribution through Z_A is greater than that through Z_D. If Z_D is very large, irrespective of the other impedances, the currents will be as shown in Z_T and Z_C. Where a tie exists, such as Z_C, outfeed is a very real possibility.

Outfeed poses significant difficulties for most relaying systems. There is the beneficial effect that tripping of source A, *reverses* the direction of an outfeed. The relay-

ing must be such that tripping at A (and B) cannot be blocked by the outfeed condition.

3. CRITERIA FOR RELAYING

The criteria for any acceptable relaying system are the following:

1. Must clear all malfunctions at the highest speed commensurate with severity of the malfunction.
2. Cost of the protective apparatus should be substantially less than the savings anticipated by its successful performance.
3. Must sensitively detected all faults and abnormal conditions in the apparatus or circuit for which it is installed to detect.
4. Must initiate isolation of a fault by de-energizing as little sound equipment and as few circuits as possible.
5. Must not respond to any normal phenomenon (such as transformer inrush current or motor-starting current).

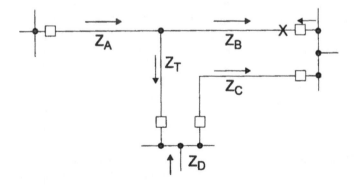

Figure 5-3 Outfeed effect.

6. Must prohibit equipment from being subjected to any stress beyond its capability (such as out-of-phase energization of rotating machinery or prolonged overcurrent consequent to a fault in another system segment).
7. Must provide backup of, at least, any single-component failure.

These constraints apply to three-terminal as well as two-terminal applications. The three-terminal applications probably require a greater amount of study to establish a workable relaying complement than do the two-terminal cases.

4. INFLUENCE OF THE LOAD AT THE TAP

With the wide variation of connections and system elements, it is evident that the protection at the third terminal may consist of anything from simple overcurrent relays to a full complement of pilot relays.

One significant factor that will influence this choice is the nature of the load at the tap. If it is a major industrial plant, equipped with its own generation, high-speed separation is dictated in the interests of minimizing the shock to the generators and to the industrial process (paper, textiles, chemicals, or other). Two possibilities exist for accomplishing this: (a) utilize a nonpilot relaying scheme and risk an occasional undesired separation for faults occurring beyond the utility supply bus, or (b) use a pilot scheme and provide high-speed tripping for all supply circuit faults and ignore all faults not on the supply circuit.

Whether a nonpilot scheme can be used or not is heavily influenced by the number of supply circuits and the amount of purchased power in relation to the magnitude of interruptible load in the plant that may be dropped without jeopardizing production. If, for example, the interruptible load exceeds the normal purchased power, an occasional unnecessary source breaker trip is an acceptable compromise in the interests of retaining the simplest tie circuit relaying. However, if the purchased power is in excess of the amount of load that can be dropped without affecting production, and if a single source breaker exists, it is not very difficult to show economic justification for the more elaborate relaying system that would prevent source breaker tripping for faults outside of the pilot relaying protective zone.

5. THREE-TERMINAL RELAYING

Some of the prevalent relaying systems that are used for three-terminal lines are described in the following.

5.1 Nonpilot Schemes

Simple short-time overcurrent relays have been used in industrial tie circuits, but unless they are torque-controlled by directional sensing, they must be coordinated with the plant relays to avoid operation for feeder faults. A far more effective scheme is to use phase and ground distance relays that have an inherent directional sense and a distinctive reach cutoff. This type of relay can operate in one to three cycles for all supply circuit faults (but also undesirably for faults somewhat beyond the supply buses as described). With the plant generation shutdown and only a small amount of motor and lighting load, separation can be accomplished by underfrequency and undervoltage relays after the supply breakers trip for a fault.

Viewing Figure 5-1, it can be seen that some of the major considerations are

1. Impedances
   a. Between A and the tap
   b. Between B and the tap
   c. Between C and the tap
2. Transformer
   a. Does it exist?
   b. How are the windings connected?
   c. What is its impedance?
3. Generation
   a. Does it exist?
   b. What is its impedance?

The impedances are important in that they help provide fault location discrimination based on current magnitude. For overcurrent relays the only ingredients available for a decision are current magnitude and time. Directional supervision adds an additional layer of discrimination, making the overcurrent relay totally inoperative unless the relative current is in the proper relation to current flow to the fault.

In the usual case, if the transformer is present, its impedance provides a discontinuity in current magnitude versus distance, allowing relays at A and B to easily recognize faults on the high-voltage side and ignore those on the low-voltage side of the transformer. In this circumstance, good use may be made of simple instantaneous overcurrent relays at A and B provided faults at X4 and behind A allow it.

Distance relays have the quality of reaching a certain distance during faults, by comparing the current at the relay with the voltage drop produced by that current. This relationship is upset when there is infeed or outfeed (see Figs. 5-2 and 5-3). This is further complicated for ground distance relays by the zero-sequence contribution from the wye winding of a wye–delta transformer at the tap, *irrespective of whether or not there is generation at the tap*.

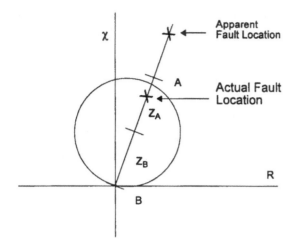

Figure 5-4 Distance relay underreach caused by infeed.

When we observe Figure 5-2 it can be seen for a 3ϕ fault at X1, that the voltage at B is $I_B Z_B + (I_B + I_C)Z_A$. The current at B is I_B. A distance relay at A examines the ratio of voltage to current and will interpret the impedances to the fault to be $Z_B + Z_A + I_C Z_A/I_B$, whereas in actuality the impedance to the fault is only $Z_B + Z_A$. This is illustrated in Figure 5-4. The relay at B will underreach by an amount equal to $I_C Z_A/I_B$ for this fault. Similarly, a distance relay applied at station C will experience underreach for faults on the line as a result of contributions from A or

B. For an outfeed condition, as described in Figure 5-3, the effect is opposite that for an infeed condition, and the setting of the distance relays must reflect this appearance of a fault being closer than it actually is.

5.2 Pilot Schemes

Most of the variations in pilot relaying that are described in Chapters 2 and 3 are applicable here. The first choice, when distances permit, is current differential. The pilot-wire type of current differential is described in Figure 5-5. Location of the tap, generation on or off, unequal current distribution factors, high-magnitude fault resistance make no difference. This scheme responds to total internal fault current. External faults, load, or power system instability do not affect this relaying system. High-speed tripping takes place at all three terminals, even though the generation is shut down for any fault on the high-voltage circuit. Other schemes that could be considered are

1. Phase comparison
2. Directional comparison
   a. Blocking
   b. Unblocking
   c. Overreaching transfer–trip
3. Underreaching transfer–trip
   a. Direct
   b. Permissive

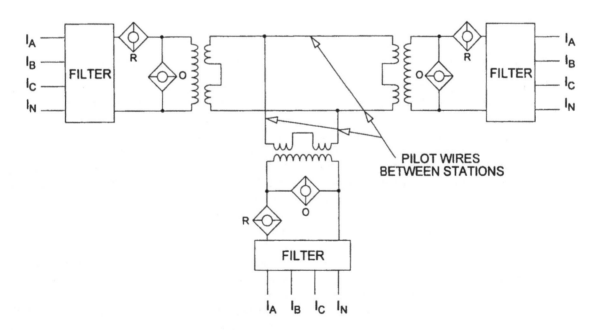

Figure 5-5 Basic pilot wire relaying system.

5.2.1 Phase Comparison

Phase-comparison systems are applicable to three-terminal line protection, but care must be exercised, particularly with an on–off carrier, in choosing the settings properly. A fault detector is used to establish the "arming" level at which tripping can take place if the phase comparison of the currents from the three terminals of the line indicates that there is an internal fault. The fault detector assures that the low-level current conditions produced by the distributed capacitance of the line cannot incorrectly cause the relays to identify a "through" condition as an internal fault. When using power line carrier, there must be a certainty that carrier can be keyed at the remote stations at any time that a local fault detector has a high enough current to operate. This requires a knowledge of the keying level, or for an on–off carrier, a critical setting of a low-set fault detector.

If we consider the possibility that there may be equal current outfeed from two terminals for an external fault, this gives a current at one terminal that is twice the level of the other two. By allowing a margin, then, the "arming" fault detector must be set at 2.5 times the setting of the "start" fault detector at all terminals. If carrier starting on load current is not allowed, this places 3ϕ-fault sensitivity at a fairly high level. The obvious conclusion from this is that phase comparison relaying in three-terminal applications, is not very sensitive from a phase fault detection standpoint. It can be very sensitive from a ground fault viewpoint, however.

Phase comparison cannot be used for protecting a three-terminal line for which outfeed may occur for an internal fault.

5.2.2 Directional Comparison

All of the directional comparison relaying systems utilize overreaching distance units. The reach settings must take into consideration the infeed effect. However, the substantial overreach that results when one terminal is open or has generation shutdown is not detrimental to the pilot system behavior. If the same distance unit drives a timer for zone-2 backup function, coordination may not be possible with adjacent line zone-1 relays. If all transmission line relaying on the power system is homogeneous, overreaching of an adjacent line zone-1 relay is a problem only if the adjacent line pilot system is out of service when a fault occurs slightly beyond the zone-1 reach. There is the possibility of coordinating the settings for adjacent zone-2 timers, but this is a painstaking, exacting process that should be avoided, if possible.

When there is no generation at the tap and no need to trip at that location, a blocking-only terminal may be set up. No action is required at the tap for an internal fault on the protected line. Blocking carrier is transmitted for external faults through the operation of a "reverse-looking" distance unit for phase faults. For external ground faults, either a directional carrier start relay (looking away from the protected line) must be used, or the nature of the tap must be such that zero-sequence current flows in the relays only for external faults (tapped delta–wye transformer).

If the transformer is treated as part of the protected zone for the line relays, and it is connected delta–wye, no ground carrier start relay is required at that location. No external fault produces zero-sequence current in the relays at that location. If the transformer is excluded from the protected zone, and a high-voltage breaker is provided, the ground carrier start function is necessary to prevent remote line breaker tripping for transformer faults. Under no circumstance should it be assumed that the line relaying will provide adequate sensitive protection for the transformer. Differential or other protection is required, with the line protection affording some degree of backup. If tripping must be accomplished at the tap for a fault on the supply circuit, a full complement of relays must be provided there.

For permissive overreaching transfer–trip (POTT; or unblocking) schemes, all terminals must recognize an internal fault. An echo scheme allows a received trip signal to cause transmission from a weak terminal, provided reverse-looking relays *do not* operate. Tripping at the weak terminal can also be accomplished through logic that identifies phase undervoltage (or zero-sequence overvoltage) plus *nonoperation* of units applied to detect external faults.

5.2.3 Underreaching Transfer-Trip

Inherent in this type of system is the use of zone-1 phase and ground distance units. In the direct-underreaching relaying system, at least one terminal must be able to reach all faults, and yet not overreach any remote bus. If the tap point on the protected line is very close to one of the other terminals and has no transformer to be included in the protected zone, this scheme is not useable. If there is a transformer in the tap that has an impedance equal to or greater than the line impedance, the scheme is useable. The zone-1 unit at the tapped terminal cannot be set, and will not be used, but the zone-1 relays at the other two terminals can be, and one or the other will reach all faults. This will initiate local trip and channel keying to produce tripping at all other terminals.

For taps closer to the center of the line, faults at the tap can be reached by one or the other of the zone-1 relays at the three stations. These settings are realizeable only if there is some significant impedance between the third terminal bus and the tap.

Operation of any zone-1 relay initiates the transmission of a channel trip as well as tripping the local breaker. All terminals may be equipped with zone-2 supervision of channel trip to add security to the system, making it a permissive underreaching transfer–trip system.

Of all the systems discussed, only current differential and direct underreaching transfer–trip may be used when an outfeed condition may exist for an internal fault.

5.2.4 Pilot Relaying Without Relaying at the Tap

In addition to the direct underreaching relaying system which *demands* tripping at the weak terminal rather than *requesting* it (which can be the strategy when a transformer exists between the tap and a feeder bus), a current differential scheme may also be useable. The transformer is a key ingredient. The available fault currents on the pro-

tected line must be substantially higher than those resulting from faults on the other side of the intervening transformer. Settings may be chosen to provide sensitive detection of all faults on the protected line and some well into the transformer, while ignoring all faults on the other side of the transformer. This does not relieve the necessity of providing independent protection for the transformer itself.

6. CONCLUSIONS

With the wide variation in system configurations and constraints, it is impossible to generalize on the proper choice of a relaying system for a three-terminal line. If the distances permit, the clear technical choice is current differential. All of the available relaying systems are applicable to certain configurations, but each power system arrangement is unique, and the relaying system must be selected with a full understanding of its limitations.

6

Program Design for Microprocessor Relays

LIANCHENG WANG

1. INTRODUCTION

Microprocessor-based relays perform protective functions by executing specifically designed computer programs. By means of these programs, various relaying algorithms may be implemented. With use of a distance relay as an example, this chapter discusses in detail relaying algorithms. Although described in the context of distance relays, the concept is equally applicable to other kinds of relay units.

Distance relays are very popular devices for line protection. A distance relay makes a trip decision based on the distance from the fault location to the relay location. If the fault distance is less than the distance setting, the fault is considered as within the protected zone. In microprocessor relaying, a distance relay can be implemented by explicitly finding the fault distance, in terms of fault resistance and fault reactance, and then comparing the computed distance with the distance setting. It also can be implemented by comparing two sinusoidal voltage signals and thus implicitly comparing the fault distance with the distance setting. In the implicit distance comparison method, either the magnitudes or phase relation (leading or lagging) of two sinusoidal signals are compared by a magnitude comparator or a phase comparator. However, no matter what method is selected, the essential task in microprocessor relaying algorithms is to create an orthogonal signal pair with both signals in the pair being 90° apart in phase.

We begin this chapter by describing orthogonal filter pairs (OFP), which generate orthogonal signals to be used in microprocessor relaying. The frequency responses of the OFPs are studied to show their characteristics under signals (noise) other than the fundamental frequency. This is followed by a parallel treatment of implicit distance relay algorithms in Section 3 and explicit algorithms in Section 4. Because of their importance in relay designs, we also introduce filters for extracting symmetrical components.

2. ORTHOGONAL SIGNALS

2.1 Orthogonal Signals in Distance Relays

Two sinusoidal signals that are 90° apart in phase are orthogonal to each other. For example, the continuous time domain signals $A \cos(\omega t + \phi)$ and $A \cos(\omega t + \phi + 90°)$ are orthogonal, where ω represent signal frequency, A represents signal magnitude, and ϕ represents phase angle. Similarly, signals $A \sin(\omega t + \phi)$ and $A \sin(\omega t + \phi + 90°)$ are orthogonal to each other. To show how orthogonal signals are applied in a protective relay, let us study a typical distance relay the inputs of which are expressed in phasor form as

$$\bar{S}_1 = Z_R \bar{I} - \bar{V} \qquad (6\text{-}1a)$$
$$\bar{S}_2 = \bar{V} \qquad (6\text{-}1b)$$

where $\bar{V}$ and $\bar{I}$ represent the voltage and current phasors seen by the relay, and Z_R is the relay reach or setting. The distance relay unit will initiate a trip signal if the angle between $\bar{S}_1$ and $\bar{S}_2$ is less than 90°. With Z representing the impedance seen by the relay, the conductance (mho) characteristics of the distance relay is a circle and illustrated in Figure 6-1. Inside the circle is the relay trip zone.

In the continuous-time domain, the relay inputs described by Eqs. (6-1a) and (6-1b) can be written as

$$s_1(t) = v(t) - R_R i(t) - \frac{X_R}{\omega} \frac{di(t)}{dt} \qquad (6\text{-}2a)$$

$$s_2(t) = v(t) \qquad (6\text{-}2b)$$

and, in the discrete time domain,

$$s_1(k) = v(k) - R_R i(k) - X_R i_{or}(k) \qquad (6\text{-}3a)$$

$$s_2(k) = v(k) \qquad (6\text{-}3b)$$

where k represents the kth sampling instant, R_R and X_R represent relay resistance and reactance settings, the current $i_{or}(k)$ has the same magnitude as $i(k)$, but the former leads the latter by 90°. So, $i_{or}(k)$ and $i(k)$ are orthogonal to each other.

The foregoing discussions clearly show that the formulation of the relay input voltage signals needs orthogonal current signal pairs. Moreover, to determine the phase relationship between $s_1(k)$ and $s_2(k)$, $s_{1,or}(k)$ and $s_{2,or}(k)$, which are orthogonal to $s_1(k)$ and $s_2(k)$ respectively, are also needed. It should be pointed out that $s_{1,or}(k)$ and $s_{2,or}(k)$ can be directly derived from $i_{or}(k)$ and $v_{or}(k)$, which is orthogonal to $v(k)$, because advancing a signal's

phase 90° twice is equivalent to multiplying it by -1. Thus, according to Eqs. (6-3a) and (6-3b),

$$s_{1,or}(k) = v_{or}(k) - R_R i_{or}(k) + X_R i(k) \qquad (6\text{-}4a)$$

$$s_{2,or}(k) = v_{or}(k) \qquad (6\text{-}4b)$$

Once we have $s_1(k)$, $s_2(k)$, as well as their orthogonal counterparts $s_{1,or}(k)$ and $s_{2,or}(k)$, the phase relationship between $s_1(k)$ and $s_2(k)$ can be identified straightforwardly. To show the procedure, let us assume

$$s_1(k) = S_1 \sin(\omega kT + \phi_1)$$

$$s_2(k) = S_2 \sin(\omega kT + \phi_2)$$

then, their orthogonal counterparts $s_{1,or}(k)$ and $s_{2,or}(k)$ are described as

$$s_{1,or}(k) = S_1 \cos(\omega kT + \phi_1)$$

$$s_{2,or}(k) = S_2 \cos(\omega kT + \phi_2)$$

Based on the basic triangular identities, we have

$$s_{1,or}(k)s_{2,or}(k) + s_1(k)s_2(k) = S_1 S_2 \cos(\phi_1 - \phi_2)$$

Therefore, whether the angle between $s_1(k)$ and $s_2(k)$ is less than 90°, or whether the relay should trip, can be determined by the sign or polarity of this expression.

2.2 Orthogonal Filter Pairs

Orthogonal signals can be generated by using orthogonal filter pairs (OFPs). Simply speaking, OFPs are two mathematical algorithms that provide orthogonal signals. In protective relay applications, the OFPs are usually linear filters. Figure 6-2 shows the OFP, represented by the discrete-time domain transfer functions $G_1(z)$ and $G_2(z)$, with the associated orthogonal signals $y(k)$ and $y_{or}(k)$.

For any sinusoidal inputs to the OFPs, their steady-state outputs are also sinusoids that have the same frequency as the inputs, but different magnitudes and phase angles. Consider a sinusoidal signal:

$$x(k) = A \sin(\omega kT + \phi) \qquad (6\text{-}5)$$

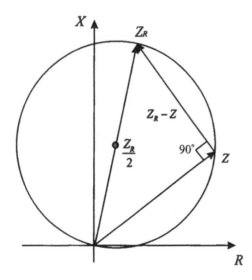

Figure 6-1 Relay mho characteristics.

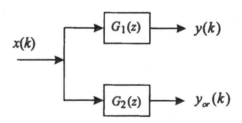

Figure 6-2 Orthogonal signals.

where k represents the kth sampling instant, T represents sampling period, A represents signal magnitude, ω represents the signal angular frequency, and ϕ represents the phase angle. With (1-1) as the input, the steady-state outputs from the OFPs can be described as

$$y(k) = AG_1 \sin(\omega kT + \phi + \theta_1) \tag{6-6a}$$

$$y_{or}(k) = AG_2 \sin(\omega kT + \phi + \theta_2) \tag{6-6b}$$

where G_1 and G_2 represent gains, θ_1 and θ_2 represent phase shifts introduced by the OFP. The gain G_1 is the magnitude of the complex quantity $G_1(e^{j\omega T})$ which is obtained by substituting $e^{j\omega T}$ for z in the transfer function $G_1(z)$, the gain G_2 is the magnitude of the complex quantity $G_2(e^{j\omega T})$, which is obtained by substituting $e^{j\omega T}$ for z in the transfer function $G_2(z)$, the angle θ_1 is the phase angle of the complex quantity $G_1(e^{j\omega T})$, and the angle θ_2 is the phase angle of the complex quantity $G_2(e^{j\omega T})$. That is,

$$G_1 = |G_1(e^{j\omega T})|, \qquad G_2 = |G_2(e^{j\omega T})|$$
$$\theta_1 = \angle G_1(e^{j\omega T}), \qquad \theta_2 = \angle G_2(e^{j\omega T})$$

Equation (6-6a) shows that $y(k)$ has the same frequency as $x(k)$ does, but its amplitude is amplified by G_1, and its phase angle is advanced by θ_1; similarly, Eq. (6-6b) shows that $y_{or}(k)$ has the same frequency as $x(k)$ does, but its amplitude is amplified by G_2 and its phase angle is advanced by θ_2. The gains G_1 and G_2, phase shifts θ_1 and θ_2 are all functions of signal frequency.

The OFP should be designed such that, at the fundamental frequency ω_0, $y_{or}(k)$, and $y(k)$ are orthogonal: $y_{or}(k)$ and $y(k)$ have the same magnitudes, but the former leads the latter by 90°. Thus, in OFP, when $\omega = \omega_0$, the following equations should hold

$$G_1 = G_2 = 1, \qquad \theta_2 = \theta_1 + 90°$$

that is, when the input is a fundamental frequency sinusoid signal

$$x(k) = A \sin(\omega_0 kT + \phi)$$

the outputs of the OFP can be described as

$$y(k) = A \sin(\omega_0 kT + \phi_1)$$
$$y_{or}(k) = A \cos(\omega_0 kT + \phi_1)$$

There exist many different ways for the designs of OFPs. Based on the characteristics of their response to impulse signals, the OFPs are classified as finite impulse response (FIR) OFPs and infinite impulse response (IIR) OFPs. The FIR OFPs are further classified as short-data window (two-samples, three-samples, and quarter-cycle)

algorithms, and long-data (one cycle and half-cycle) algorithms. This section describes typical OFPs under each category of the FIR OFPs along with their frequency responses. In showing the frequency responses, the sampling rate (samples per fundamental frequency cycle) is assumed to be 12.

2.2.1 Short-Data Window Orthogonal Filter Pairs

Short-data window OFPs utilize two- or three-consecutive samples, usually the current sample plus the previous one or two samples. Because they do not need long memory, these OFPs are easier to implement, but have poor immunity to high-frequency noise. Thus, they should be used in combinations with other filters.

Two-Sample Orthogonal Filter Pairs

Algorithm 1 In this two-sample OFP, one filter directly takes its input as its output (no filtering), while the other filter shifts its input 90° for fundamental frequency signals. In terms of the input series $x(k)$, the OFPs outputs $y(k)$ and $y_{or}(k)$ are described as

$$y(k) = x(k) \tag{6-7a}$$
$$y_{or}(k) = c_0 x(k) - c_1 x(k - 1) \tag{6-7b}$$

where c_0 and c_1 are sampling frequency-dependent constants, and

$$c_0 = \frac{\cos \omega_0 T}{\sin \omega_0 T}$$

$$c_1 = \frac{1}{\sin \omega_0 T}$$

These constants are used to ensure that $y(k)$ and $y_{or}(k)$ are orthogonal to each other at the fundamental frequency. By taking the z transform of Eqs. (6-7a) and (6-7b), the transfer functions of the two-sample OFP can be described as

$$G_1(z) = 1 \tag{6-8a}$$

$$G_2(z) = \frac{1}{\sin \omega_0 T}(\cos \omega_0 T - z^{-1}) \tag{6-8b}$$

where z^{-1} represents one-sample time delay. Based on Eqs. (6-8a) and (6-8b), the block diagram of the two sample OFP is shown in Figure 6-3.

To find the frequency response, substituting $e^{j\omega T}$ in Eqs. (6-8a) and (6-8b) for z yields

$$G_1(e^{j\omega T}) = 1 \tag{6-9a}$$

$$G_2(e^{j\omega T}) = G_2 e^{j\theta_2} \tag{6-9b}$$

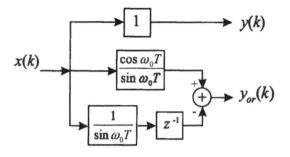

Figure 6-3 Block diagram of two-sample OFP (see algorithm 1).

where

$$G_2 = \frac{(1 + \cos^2 \omega_0 T - 2 \cos \omega_0 T \cos \omega T)^{1/2}}{\sin \omega_0 T}$$

$$\theta_2 = \frac{\sin \omega T}{\cos \omega_0 T - \cos \omega T}$$

Obviously, at the fundamental frequency, the magnitude of $G_2(z)$ is equal to 1 and the angle shift by $G_2(z)$ is 90°; thus, $y(k)$ and $y_{or}(k)$ are orthogonal. The magnitude responses of $G_1(z)$ and $G_2(z)$ are shown in Figure 6-4. Filter $G_2(z)$ (first derivative) attenuates low-frequency signals, but amplifies high-frequency signals. The angle shift between $y_{or}(k)$ and $y(k)$, which is introduced by $G_1(z)$ and $G_2(z)$, is shown in Figure 6-5. The angle shift decreases with the increase of signal frequencies and is equal to 90° at the fundamental frequency and 0 at the cutoff frequency.

Algorithm 2 In the OFP, one filter takes the arithmetic average between the most recent two samples and multiplies the average by a constant, which is determined by the sampling frequency. This constant is used to compensate magnitude attenuation for fundamental frequency signals. The other filter takes the difference (first derivative) between the most recent two samples and modifies the result by another constant. Also, this constant is determined by the sampling frequency and used for compensating fundamental frequency signal attenuation. The outputs of the OFP are described as

$$y(k) = \frac{1}{2 \cos(\omega_0 T/2)}[x(k) + x(k-1)] \quad (6\text{-}10a)$$

$$y_{or}(k) = \frac{1}{2 \sin(\omega_0 T/2)}[x(k) - x(k-1)] \quad (6\text{-}10b)$$

By taking z transform of Eqs. (6-10a) and (6-10b), the transfer functions of the two-sample OFP can be described as

$$G_1(z) = \frac{1}{2 \cos(\omega_0 T/2)}(1 + z^{-1}) \quad (6\text{-}11a)$$

$$G_2(z) = \frac{1}{2 \sin(\omega_0 T/2)}(1 - z^{-1}) \quad (6\text{-}11b)$$

From Eqs. (6-11a) and (6-11b), the block diagram of the two sample OFP is shown in Figure 6-6.
To study the frequency response of the OFP, substituting $e^{j\omega T}$ in Eqs. (6-11a) and (6-11b) for z yields

$$G_1(e^{j\omega T}) = \frac{1 + e^{-j\omega T}}{2 \cos(\omega_0 T/2)} = \frac{\cos(\omega T/2)}{\cos(\omega_0 T/2)}e^{-j\frac{\omega T}{2}} \quad (6\text{-}12a)$$

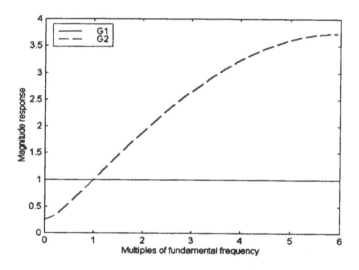

Figure 6-4 Magnitude responses of two-sample OFP (see algorithm 1).

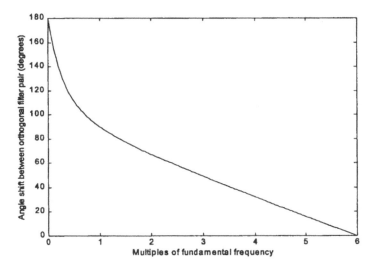

Figure 6-5 Phase shift between two-sample OFP (see algorithm 1).

$$G_2(e^{j\omega T}) = \frac{1 - e^{-j\omega T}}{2\sin(\omega_0 T/2)} \quad (6\text{-}12b)$$

$$= \frac{\sin(\omega T/2)}{\sin(\omega_0 T/2)} e^{j(\frac{\pi}{2} - \frac{\omega T}{2})}$$

It can be easily verified that, at the fundamental frequency ω_0, the gains of both filters in the OFP are equal to 1:

$$|G_1(e^{j\omega_0 T})| = |G_2(e^{j\omega_0 T})| = 1$$

The magnitude response of $G_1(z)$ and $G_2(z)$ are shown in Figure 6-7. The angle shift between $G_1(z)$ and $G_2(z)$ is 90° for all signal frequencies.

Three-Sample Orthogonal Filter Pairs

Three-sample OFPs have a memory of the most recent three samples. Therefore, as with the two-sample OFPs,

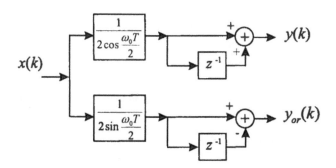

Figure 6-6 Block diagram of two-sample OFP (see algorithm 2).

their response time is very fast—three samples. They have the same disadvantage as the two-sample OFPs (i.e., poor response to high-frequency noise).

Algorithm 1 In this three-sample OFP, one filter takes the sample before the current one as its output, and the other filter takes the difference (first derivative) between the current sample and the third sample counting from the current one backward. The second filter is modified by a constant to compensate the magnitude attenuation of the fundamental frequency sinusoid. In the discrete time domain, the OFP outputs $y(k)$ and $y_{or}(k)$ are described as

$$y(k) = x(k - 1) \quad (6\text{-}13a)$$

$$y_{or}(k) = \frac{1}{2\sin\omega_0 T}[x(k) - x(k - 2)] \quad (6\text{-}13b)$$

By taking the z transform of Eqs. (6-13a) and (6-13b), the transfer functions of the two-sample OFP are described as

$$G_1(z) = z^{-1} \quad (6\text{-}14a)$$

$$G_2(z) = \frac{1}{2\sin\omega_0 T}(1 - z^{-2}) \quad (6\text{-}14b)$$

where z^{-2} represents two-sample time delay. Based on Eqs. (6-14a) and (6-14b), the block diagram of the three-sample OFP is shown in Figure 6-8.

Substituting $e^{j\omega T}$ for z in Eqs. (6-14a) and (6-14b) yields

$$G_1(e^{j\omega T}) = e^{-j\omega T} \quad (6\text{-}15a)$$

$$G_2(e^{j\omega T}) = \frac{1 - e^{-j2\omega T}}{2\sin\omega_0 T} = \frac{\sin\omega T}{\sin\omega_0 T} e^{j(\frac{\pi}{2} - \omega T)} \quad (6\text{-}15b)$$

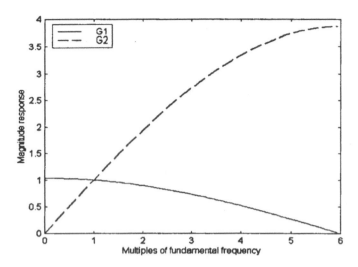

Figure 6-7 Magnitude responses of two-sample OFP (see algorithm 2).

It can be verified that, at the fundamental frequency ω_0, the gains of both filters in the OFP are equal to 1:

$$|G_1(e^{j\omega_0 T})| = |G_2(e^{j\omega_0 T})| = 1 \tag{6-16}$$

The magnitude response of $G_1(z)$ and $G_2(z)$ are shown in Figure 6-9. The angle shift between $G_1(z)$ and $G_2(z)$ is 90° for all signal frequencies.

Algorithm 2 In this three-sample OFP, one filter takes the difference (first derivative) between the current sample and the third sample counting from the current one backward, and the other filter takes a weighted sum (second derivative) of the most recent three samples. The weighting factors for the three samples are 1, −2, and 1, respectively. Both filters are modified by constants to compensate the magnitude attenuation for fundamental frequency sinusoids. In the discrete-time domain, the OFP outputs $y(k)$ and $y_{or}(k)$ are described as

$$y(k) = \frac{1}{2 \sin \omega_0 T}[x(k) - x(k - 2)] \tag{6-17a}$$

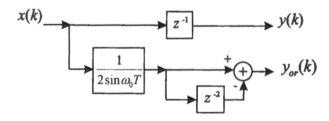

Figure 6-8 Block diagram of three-sample OFP (see algorithm 1).

$$y_{or}(k) = \frac{1}{4 \sin^2(\omega_0 T/2)}[x(k) - 2x(k - 1) + x(k - 2)] \tag{6-17b}$$

If we take z transform of Eqs. (6-17a) and (6-17b), the transfer functions for the three-sample OFP can be expressed as

$$G_1(z) = \frac{1}{2 \sin \omega_0 T}(1 - z^{-2}) \tag{6-18a}$$

$$G_2(z) = \frac{1}{4 \sin^2(\omega_0 T/2)}(1 - z^{-1})^2 \tag{6-18b}$$

From Eqs. (6-18a) and (6-18b), the block diagram of the three-sample OFP is shown in Figure 6-10.

Substitution of $e^{j\omega T}$ for z in Eqs. (6-18a) and (6-18b) yields

$$G_1(e^{j\omega T}) = \frac{1 - e^{-j2\omega T}}{2 \sin \omega_0 T} = \frac{\sin \omega T}{\sin \omega_0 T}e^{j(\pi/2 - \omega T)} \tag{6-19a}$$

$$G_2(e^{j\omega T}) = \frac{(1 - e^{-j\omega T})^2}{4 \sin^2(\omega_0 T/2)} \tag{6-19b}$$

$$= \frac{\sin^2(\omega T/2)}{\sin^2(\omega_0 T/2)}e^{j(\pi - \omega T)}$$

At the fundamental frequency ω_0, the gains of the two filters are equal to 1:

$$|G_1(e^{j\omega_0 T})| = |G_2(e^{j\omega_0 T})| = 1 \tag{6-20}$$

The magnitude response of $G_1(z)$ and $G_2(z)$ are shown in Figure 6-11. The angle shift between $G_1(z)$ and $G_2(z)$ is 90° for all signal frequencies.

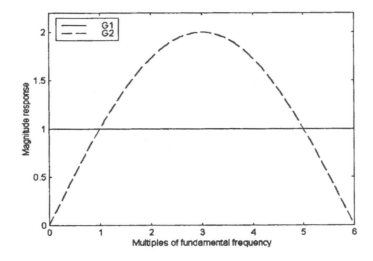

Figure 6-9 Magnitude responses of three-sample OFP (see algorithm 1).

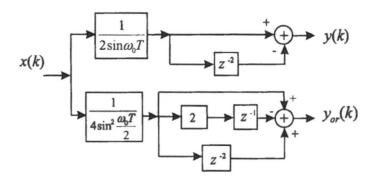

Figure 6-10 Block diagram of three-sample OFP (see algorithm 2).

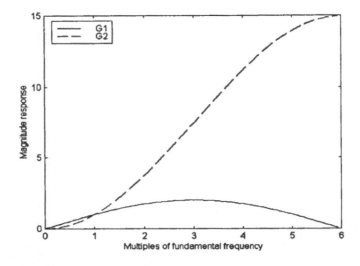

Figure 6-11 Magnitude responses of three-sample OFP (see algorithm 2).

Quarter-Cycle Delaying Orthogonal Filter Pair

The quarter-cycle delaying OFPs use time delaying to obtain orthogonal signals. The required number of delayed samples depends on sampling rate and should be equivalent to a quarter of fundamental frequency cycle. The outputs of the quarter-cycle OFP are expressed as

$$y(k) = x\left(k - \frac{N}{4}\right) \tag{6-21a}$$

$$y_{or}(k) = x(k) \tag{6-21b}$$

Thus, the transfer functions of the OFP can be written as

$$G_1(z) = z^{-(N/4)} \tag{6-22a}$$

$$G_2(z) = 1 \tag{6-22b}$$

where $z^{(-N/4)}$ represents quarter-cycle time delay. Based on Eqs. (6-22a) and (6-22b), the block diagram of the quarter-cycle OFP is shown in Figure 6-12.

To show the frequency response of the quarter-cycle delay OFP, substituting $e^{j\omega T}$ for z in Eqs. (6-22a) and (6-22b) yields

$$G_1(e^{j\omega T}) = e^{-j(N/4)\omega T} \tag{6-23a}$$

$$G_2(e^{j\omega T}) = 1 \tag{6-23b}$$

Obviously, the gains of $G_1(e^{j\omega T})$ and $G_2(e^{j\omega T})$ are the same for all signal frequencies. The angle shift between $G_1(z)$ and $G_2(z)$ is shown in Figure 6-13. At the fundamental frequency ω_0,

$$\angle G_2(e^{j\omega_0 T}) - \angle G_1(e^{j\omega_0 T}) = \frac{\pi}{2} \tag{6-24}$$

2.2.2 Long-Data Window Orthogonal Filter Pairs

Long-data window OFPs utilize the most recent one or half cycle's data in obtaining orthogonal signals. Thus, the response speed of long-data window OFPs is slower than that of the short-data window OFPs. The advantage of long-data window OFPs is their excellent immune capacity to high-frequency noise. This section presents two kinds of long-data window OFPs: full-cycle DFT and half-cycle DFT.

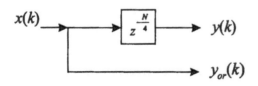

Figure 6-12 Block diagram of quarter-cycle OFP.

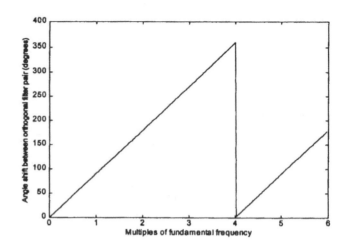

Figure 6-13 Phase shift between quarter-cycle delaying OFP.

Full-Cycle DFT

The full-cycle DFT OFP take weighted averages of samples within the most recent one cycle. The weighting factors are cosine and sine of angles taken within the range of 2π with the increment of $2\pi/N$. In terms of their input series $x(k)$, the OFP outputs are described as

$$y(k) = \frac{2}{N}\left[\cos\left(\frac{2\pi}{N}0\right)x(k) + \cos\left(\frac{2\pi}{N}\right)x(k-1)\right.$$
$$+ \cos\left(\frac{2\pi}{N}2\right)x(k-2) + \cdots \tag{6-25a}$$
$$\left. + \cos\left(2\pi - \frac{2\pi}{N}\right)x(k-N+1)\right]$$

$$y_\perp(k) = -\frac{2}{N}\left[\sin\left(\frac{2\pi}{N}0\right)x(k) + \sin\left(\frac{2\pi}{N}\right)x(k-1)\right.$$
$$+ \sin\left(\frac{2\pi}{N}2\right)x(k-2) + \cdots \tag{6-25b}$$
$$\left. + \sin\left(2\pi - \frac{2\pi}{N}\right)x(k-N+1)\right]$$

By taking the z transform of Eqs. (6-25a) and (6-25b), the transfer functions of the full-cycle DFT OFP can be described as

$$G_1(z) = \frac{2}{N}\sum_{i=0}^{N-1}\cos\left(\frac{2\pi}{N}i\right)z^{-1} \tag{6-26a}$$

$$G_1(z) = -\frac{2}{N}\sum_{i=0}^{N-1}\sin\left(\frac{2\pi}{N}i\right)z^{-1} \tag{6-26b}$$

From Eqs. (6-26a) and (6-26b), the block diagram of the full-cycle DFT OFP is shown in Figure 6-14.

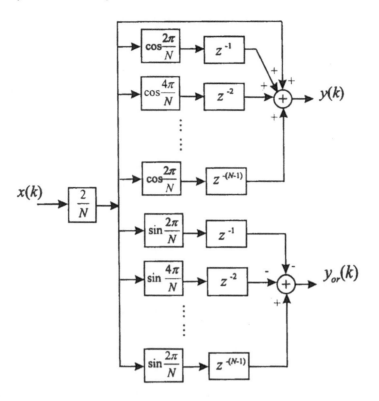

Figure 6-14 Block diagram of full-cycle DFT OFP.

The magnitude responses of the full-cycle DFT OFP are shown in Figure 6-15. It can be seen that the OFP can eliminate constant dc and integer harmonic components. They can also greatly reduce the effect of noninteger harmonic components The phase shift between the OFP is shown in Figure 6-16. At the fundamental frequency, the OFP have the same gains and provide the desired 90°

phase shift. The OFP are orthogonal only at the fundamental frequency.

Half-Cycle DFT

The half-cycle DFT OFP take weighted averages of samples within the most recent half a cycle. The weighting

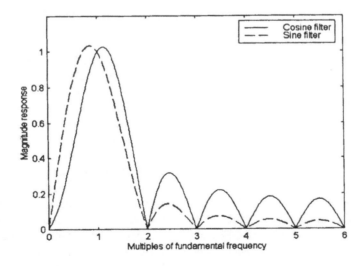

Figure 6-15 Magnitude responses of full-cycle DFT OFP.

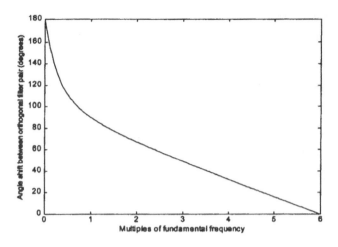

Figure 6-16 Phase shift between full-cycle DFT OFP.

factors are cosine and sine of angles taken within the range of π with the increment of $2\pi/N$. The outputs of the half-cycle OFP are described as

$$y(k) = -\frac{4}{N}\left[\cos\left(\frac{2\pi}{N}0\right)x(k) + \cos\left(\frac{2\pi}{N}\right)x(k-1)\right.$$

$$+ \cos\left(\frac{2\pi}{N}2\right)x(k-2) + \cdots \qquad (6\text{-}27a)$$

$$\left. + \cos\left(\pi - \frac{2\pi}{N}\right)x\left(k - \frac{N}{2} + 1\right)\right]$$

$$y_\perp(k) = \frac{4}{N}\left[\sin\left(\frac{2\pi}{N}0\right)x(k) + \sin\left(\frac{2\pi}{N}\right)x(k-1)\right.$$

$$+ \sin\left(\frac{2\pi}{N}2\right)x(k-2) + \cdots \qquad (6\text{-}27b)$$

$$\left. + \sin\left(\pi - \frac{2\pi}{N}\right)x\left(k - \frac{N}{2} + 1\right)\right]$$

Thus, the transfer functions of the half-cycle DFT OFP are described as

$$G_1(z) = -\frac{4}{N}\sum_{i=0}^{N/2-1}\cos\left(\frac{2\pi}{N}i\right)z^{-i} \qquad (6\text{-}28a)$$

$$G_2(z) = \frac{4}{N}\sum_{i=0}^{N/2-1}\sin\left(\frac{2\pi}{N}i\right)z^{-i} \qquad (6\text{-}28b)$$

From Eqs. (6-28a) and (6-28b), the block diagram of the half-cycle DFT OFP is shown in Figure 6-17.

The magnitude responses of the half-cycle DFT OFP are shown in Figure 6-18. It can be seen that the OFP can eliminate only odd-integer harmonic components from their inputs and have little or no attenuation on constant dc

and second-harmonic components. The phase shift between the OFP is shown in Figure 6-19. At the fundamental frequency, the OFP have the same gains and provide the desired 90° phase shift. Also, the half-cycle DFT OFP are orthogonal only at the fundamental frequency.

Note that in Figures 6-5, 6-16, and 6-19, the phase shifts in algorithm 1 of the two-sample OFPs, full-cycle DFT OFP, and half-cycle DFT OFP are exactly the same.

3. IMPLICIT (TORQUE-LIKE) RELAY ALGORITHMS

In implicit or torque-like relay algorithms, whether a fault is within a protected zone is determined by the means of torque, which may be created from either a phase comparator or magnitude comparator.

The concept of torque was first applied to designs of distance relays more than four decades ago, which resulted in the well-performed electromechanical (EM) cylinder unit. In fact, some microprocessor relays designed over a decade ago also utilized the concept of torque. Unfortunately, in these microprocessor relays, only the polarity (positive or negative) of a torque was employed, but the magnitude of a torque, which indicates how far a fault is from reach boundary, was disregarded. The resulting relay treats all faults within a protected zone blindly in the same way, and there is no intentional interference on its operating speed regardless of fault locations within the zone. Only recently has the importance of the torque magnitude in the designs of high-speed microprocessor relays been appreciated. The applications of post-filters, which take into account both the polarity and magnitude of a torque, make the distance relay have the desirable inverse characteristics: the relay takes less time to make a decision for severe close-in faults and more time for faults near reach boundary.

3.1 Phase Comparator, Magnitude Comparator, and Torque

This section describes the relation between a phase comparator and a magnitude comparator and introduces the concept of torque created by a phase comparator and a magnitude comparator. The advantage of comparing torque, rather than angles or magnitudes, in a distance relay is discussed. Finally, the section studies the condition when a phase comparator is equivalent to a magnitude comparator in the sense of torque.

Phase comparators and magnitude comparators are the basic elements in building protective relays that include EM relays, solid-state relays, and microprocessor relays.

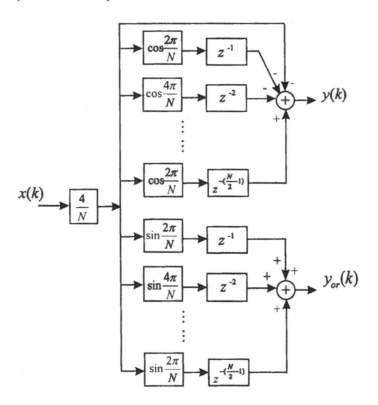

Figure 6-17 Block diagram of half-cycle DFT OFP.

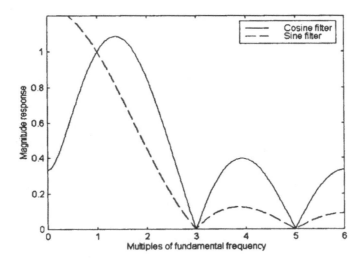

Figure 6-18 Magnitude responses of half-cycle DFT OFP.

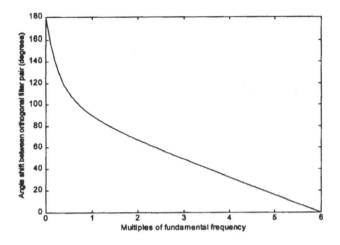

Figure 6-19 Phase shift between half-cycle DFT OFP.

Among phase comparators, two comparators are the most often used: phase comparator I and phase comparator II. For phase comparator I, the relay trip condition is that the angle between the comparator's two inputs is less than 90°:

$$-90° \le \angle\left(\frac{\overline{S}_1}{\overline{S}_2}\right) \le 90° \tag{6-29}$$

where $\overline{S}_1$ and $\overline{S}_2$ represent the phase comparator inputs.

Figure 6-20 shows the operation region (shaded area) of the relay that utilizes phase comparator I. As shown in the figure, the relay operation region and restraint region are composed of half planes divided by the line that is perpendicular to the reference phasor.

For phase comparator II, the relay trip condition is that one of its inputs leads its other input:

$$0° \le \angle\left(\frac{\overline{S}_1}{\overline{S}_2}\right) \le 180°$$

Figure 6-21 shows the operation region (shaded area) of the relay that utilizes phase comparator II. From Figure 6-21, the relay operation region and restraint region are also half planes but divided by the reference phasor. Depending on the reference phasor selected, the operation region could be on the left- or right-hand side of the reference phasor.

For the magnitude comparator, the relay trip condition is that the magnitude of one of its inputs is larger than the magnitude of its other input:

$$|\overline{S}_A| \ge |\overline{S}_B| \tag{6-30}$$

where $\overline{S}_A$ and $\overline{S}_B$ represent the magnitude comparator inputs. Figure 6-22 shows the operation region of the relay that utilizes the magnitude comparator. Note that the relay operation region and restraint region are separated by a circle, with its radius being equal to the magnitude of the reference phasor. Depending on the reference phasor selected, the operation region could be inside or outside of the circle.

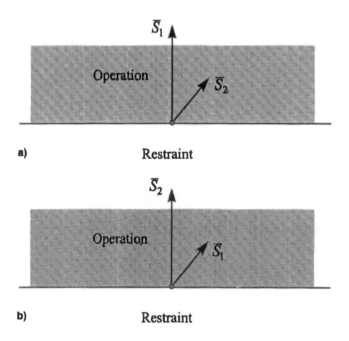

Figure 6-20 Phase comparator I: (a) $\overline{S}_1$ as reference phasor; (b) $\overline{S}_2$ as reference phasor.

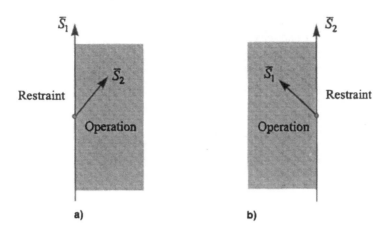

Figure 6-21   Phase comparator II: (a) $\bar{S}_1$ as reference phasor; (b) $\bar{S}_2$ as reference phasor.

Phase comparators and magnitude comparators can be used interchangeably in relay designs. We may use a magnitude comparator to compare the phase angles of two signals or use a phase comparator to compare the magnitudes of two signals. For instance, we can determine if the angle between two phasors $\bar{S}_1$ and $\bar{S}_2$ is less than 90°:

$$-90° \leq \angle\left(\frac{\bar{S}_1}{\bar{S}_2}\right) \leq 90°$$

by comparing the magnitudes of two other phasors $\bar{S}_A$ and $\bar{S}_B$:

$$|\bar{S}_A| \geq |\bar{S}_B|$$

where

$$\bar{S}_A = K(\bar{S}_1 + \bar{S}_2) \qquad\qquad (6\text{-}31a)$$
$$\bar{S}_B = K(\bar{S}_1 - \bar{S}_2) \qquad\qquad (6\text{-}31b)$$

and K is an arbitrary constant. That is, the angle between $\bar{S}_1$ and $\bar{S}_2$ being within 90° is equivalent to the magnitude of $\bar{S}_A$ being larger than the magnitude of $\bar{S}_B$:

$$-90° \leq \angle\left(\frac{\bar{S}_1}{\bar{S}_2}\right) \leq 90° \Leftrightarrow |\bar{S}_A| \geq |\bar{S}_B|$$

so long as the magnitude comparator inputs $\bar{S}_A$ and $\bar{S}_B$ are proportional to the sum and subtraction of the phasor comparator inputs $\bar{S}_1$ and $\bar{S}_2$, respectively.

To show how phase comparators and magnitude comparators are applied in distance relay units, let us return to the distance unit introduced in Section 2.1. The two inputs of the unit is given in Eq. (6-1) and repeated here for easy reference:

$$\bar{S}_1 = Z_R\bar{I} - \bar{V} = (Z_R - Z)\bar{I}$$
$$\bar{S}_2 = \bar{V} = Z\bar{I}$$

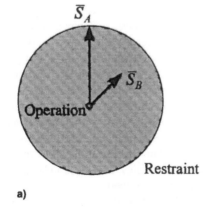

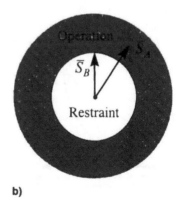

Figure 6-22   Magnitude comparator: (a) $\bar{S}_A$ as reference phasor; (b) $\bar{S}_B$ as reference phasor.

The trip condition of the unit is that the angle between $\bar{S}_1$ and $\bar{S}_2$ is less than 90°. The conductive (mho) characteristics of the distance relay is a circle and illustrated in Figure 6-23. Inside the circle is the relay trip zone. Given the relationship between phase comparator and magnitude comparator, the same relay characteristics also can be implemented by using a magnitude comparator. According to Eq. (6-31), the inputs of the magnitude comparator may be selected as

$$\bar{S}_A = \frac{Z_R}{2}\bar{I}$$

$$\bar{S}_B = \bar{V} - \frac{Z_R}{2}\bar{I} = \left(Z - \frac{Z_R}{2}\right)\bar{I}$$

The relay trip condition is $|\bar{S}_A| \geq |\bar{S}_B|$, and the resulting relay characteristics is illustrated in Figure 6-24, with the relay trip region being inside the circle.

Figure 6-23 shows that, at the *maximum torque angle*, which is defined as the angle when the fault impedance and relay setting impedance have the same angle, the angle difference ϕ between the phase comparator inputs $\bar{S}_1$ and $\bar{S}_2$ is 0° for faults inside the protected zone and 180° outside the zone:

$$\phi = \angle\bar{S}_1 - \angle\bar{S}_2 = \begin{cases} 0°, & |Z| < |Z_R| \\ 180°, & \text{otherwise} \end{cases}$$

The phase angle difference ϕ as a function of fault impedance $|Z|$ is shown in Figure 6-25.

Figure 6-24 shows that at the maximum torque angle, the magnitude difference between its inputs $\bar{S}_A$ and $\bar{S}_B$ is

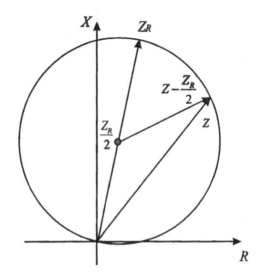

Figure 6-24 Relay mho characteristics by a magnitude comparator.

directly proportional to fault impedance when the impedance is less than half of the relay setting, and is inversely proportional to fault impedance when it is larger than half of the relay setting:

$$\Delta M = |\bar{S}_A| - |\bar{S}_B|$$

$$= \begin{cases} |Z||I|, & |Z| < \frac{|Z_R|}{2} \\ (|Z_R| - |Z|)|I|, & \text{otherwise} \end{cases}$$

The magnitude difference ΔM is illustrated in Figure 6-26. ΔM is positive inside the protected zone and negative outside the protected zone. Furthermore, the magnitude difference is maximum at the center of the relay mho circle and decreases as the fault distance becomes closer to the mho circle boundary.

From Figure 6-25 and 6-26, it can be concluded that both the phase difference in the phase comparator and the magnitude difference in the magnitude comparator are valid quantities in indicating if a fault is inside the protected zone, but neither can tell if the fault is near the zone reach boundary or is a close-in fault. In contrast, as will be shown in the next section, the torque generated by a phase or magnitude comparator is very sensitive to fault locations and, thus, can serve the latter purpose well. This makes it possible to design a fast-operating distance relay for close-in faults and with high reach accuracy as well.

According to the notation in cylinder EM relays, the torque generated by a phase comparator, the operating condition of which is defined by Eq. (6-29), is described as the

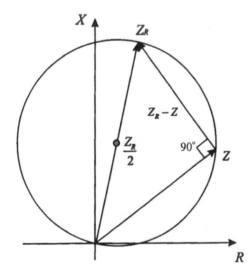

Figure 6-23 Relay mho characteristics by a phase comparator.

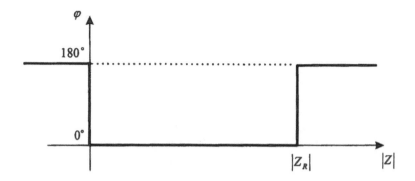

Figure 6-25 Phase angle difference between the two inputs of a phase comparator at the maximum torque angle.

sum of the real parts product and imaginary parts product of its inputs,

$$Tq_p = S_{1r}S_{2r} + S_{1i}S_{2i} = |\bar{S}_1||\bar{S}_2| \cos \phi \qquad (6\text{-}32)$$

where Tq_p is the torque generated by the phase comparator; $\bar{S}_1$ and $\bar{S}_2$ are its two input phasors; S_{1r} and S_{1i} are the real and imaginary parts of $\bar{S}_1$, respectively; S_{2r} and S_{2i} are the real and imaginary parts of $\bar{S}_2$, respectively; ϕ is the angle difference between $\bar{S}_1$ and $\bar{S}_2$. If the operating condition of a phase comparator is $\bar{S}_1$ leading $\bar{S}_2$:

$$0° \leq \angle\left(\frac{\bar{S}_1}{\bar{S}_2}\right) \leq 180° \qquad (6\text{-}33)$$

the torque generated by the comparator is described as the subtraction between the real and imaginary parts products of its inputs:

$$Tq_p = S_{1r}S_{2r} - S_{1i}S_{2i} = |\bar{S}_1||\bar{S}_2| \sin \phi \qquad (6\text{-}34)$$

Similarly, for a magnitude comparator, the operating condition of which is defined by Eq. (6-30), the torque gener-

ated by the comparator is described as the subtraction between the magnitude squares of its inputs:

$$\begin{aligned} Tq_m &= |\bar{S}_A|^2 - |\bar{S}_B|^2 = (S_{Ar}^2 + S_{Ai}^2) - (S_{Br}^2 + S_{Bi}^2) \\ &= (S_{Ar} + S_{Br})(S_{Ar} - S_{Br}) \\ &+ (S_{Ai} + S_{Bi})(S_{Ai} - S_{Bi}) \end{aligned} \qquad (6\text{-}35)$$

where Tq_m is the torque generated by the magnitude comparator; $\bar{S}_A$ and $\bar{S}_B$ are its two input phasors; S_{Ar} and S_{Ai} are the real and imaginary parts of $\bar{S}_A$, respectively; S_{Br} and S_{Bi} are the real and imaginary parts of $\bar{S}_B$, respectively. When we compare Eqs. (6-32) and (6-35), it can be found that a phase comparator is equivalent to a magnitude comparator in the sense of torque if

$$\bar{S}_1 = \bar{S}_A + \bar{S}_B \qquad (6\text{-}36a)$$
$$\bar{S}_2 = \bar{S}_A - \bar{S}_B \qquad (6\text{-}36b)$$

or

$$S_A = \frac{\bar{S}_1 + \bar{S}_2}{2} \qquad (6\text{-}37a)$$

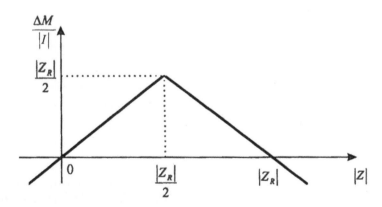

Figure 6-26 Magnitude difference between the two inputs of a magnitude comparator at the maximum torque angle.

$$\bar{S}_B = \frac{\bar{S}_1 - \bar{S}_2}{2} \tag{6-37b}$$

Therefore, when we design a distance relay based on torque, we also have the freedom to choose either a phase comparator or a magnitude comparator because they both can generate exactly the same torque if their inputs satisfy Eqs. (6-36) or (6-37). It should be pointed out that the equivalence of a phase comparator to a magnitude comparator resulting from Eq. (6-31) implies region-to-region mapping: trip or nontrip. However, the equivalence resulting from Eq. (6-37) implies point-to-point mapping.

3.2 Torque in Distance Units

This section analyzes the torque generated in ground and phase-to-phase distance units. The ground units considered are cross-polarized distance units, and the phase-to-phase unit is a polyphase distance unit. Torque for other types of distance units can be analyzed in the same way. After the torque is described in terms of source parameters and line parameters, it is normalized such that all distance units have the same torque expression, which simplifies the design of post-filters that use torque as their inputs.

3.2.1 Ground Units

Ground impedance units are designed for detecting the location of single phase-to-ground faults, including A-to-ground, B-to-ground, and C-to-ground faults. In cross-polarized ground impedance units, the operating and reference quantities may be chosen as

$$\bar{S}_1 = (\bar{I}_a + K_0\bar{I}_0)Z_R - \bar{V}_a \tag{6-38a}$$

$$\bar{S}_2 = \frac{1}{\sqrt{3}}(\bar{V}_b - \bar{V}_c) \tag{6-38b}$$

where $\bar{I}_a$ and $\bar{I}_0$ represent the phase A current and zero-sequence current, respectively; $\bar{V}_a$, $\bar{V}_b$, and $\bar{V}_c$ represent the voltages at the relay location in phases A, B, and C, respectively; Z_R is the relay setting; $K_0 = (Z_0 - Z)/Z$ representing the zero-sequence current compensation factor with Z_0 and Z being the transmission line zero-sequence and positive-sequence impedance, respectively. The relay trip condition is

$$0° \le \angle\left(\frac{\bar{S}_1}{\bar{S}_2}\right) \le 180°$$

At the relay location, the voltage in the phase A (faulted phase) is equal to the voltage drop of the phase A compen-

sated current $\bar{I}_a + K_0\bar{I}_0$ on the positive-sequence line impedance Z:

$$\bar{V}_a = (\bar{I}_a + K_0\bar{I}_0)Z \tag{6-39}$$

Substituting Eq. (6-39) into Eq. (6-38a) yields,

$$\bar{S}_1 = (\bar{I}_a + K_0\bar{I}_0)(Z_R - Z) \tag{6-40}$$

On the other hand hand, the phase A source voltage $\bar{E}_a$ is related to the fault current by

$$\bar{E}_a = (Z + Z_s)\bar{I}_a + (K_0 Z + K_{0s}Z_s)\bar{I}_0 \tag{6-41}$$

where K_{0s} represents the source zero-sequence current compensation factor. With Z_{0s} being the source zero-sequence impedance, K_{0s} is defined as

$$K_{0s} = \frac{Z_{0s} - Z_s}{Z_s}$$

Thus, if we assume the source and line zero-sequence current compensation factors are the same, the phase A compensated current can be described as the ratio between the source voltage and Thevenin's equivalent positive-sequence impedance seen at the fault location

$$\bar{I}_a + K_0\bar{I}_0 = \frac{\bar{E}_a}{Z + Z_s} \tag{6-42}$$

Substituting Eq. (6-42) into Eq. (6-40), we get

$$\bar{S}_1 = \bar{E}_a\left(\frac{Z_R - Z}{Z + Z_s}\right) \tag{6-43}$$

or equivalently,

$$\bar{S}_1 = \bar{E}_a\left[\left|\frac{Z_R + Z_s}{Z + Z_s}\right|\cos(\phi_R' - \phi') - 1 \right. \tag{6-44}$$
$$\left. + j\left|\frac{Z_R + Z_s}{Z + Z_s}\right|\sin(\phi_R' - \phi')\right]$$

where ϕ_R' is the impedance angle of $Z_R + Z_s$, ϕ' is the impedance angle of $Z + Z_s$:

$$\phi_R' = \angle(Z_R + Z_s)$$
$$\phi' = \angle(Z + Z_s)$$

Because load current is not considered, the polarizing quantity Eq. (6-38b) can be simplified as

$$\bar{S}_2 = -j\bar{E}_a \tag{6-45}$$

Substituting Eqs. (6-44) and (6-45) into Eq. (6-34), the torque generated by the ground distance unit is found as

$$Tq = S_1 S_{2r} - S_1 S_{2i}$$
$$= E_a^2 \frac{1}{|Z + Z_s|} [|Z_R + Z_s| \tag{6-46}$$
$$\cos(\phi_R' - \phi') - |Z + Z_s|]$$

If we use a magnitude comparator that creates the same torque as this phase comparator, based on Eq. (6-37), the inputs to the magnitude comparator are determined as

$$\bar{S}_A = \frac{1}{2}(\bar{S}_1 + j\bar{S}_2) \tag{6-47a}$$

$$\bar{S}_B = \frac{1}{2}(\bar{S}_1 - j\bar{S}_2) \tag{6-47b}$$

According to Eqs. (6-43) and (6-45), the operating and restraint quantities of the magnitude comparator can be simplified as

$$\bar{S}_A = \frac{1}{2}\bar{E}_a\left(\frac{Z_R - Z}{Z + Z_s} + 1\right) \tag{6-48a}$$
$$= \left(\frac{\bar{E}_a}{Z + Z_s}\right)\left(\frac{Z_R + Z_s}{2}\right)$$

$$\bar{S}_B = \frac{1}{2}\bar{E}_a\left(\frac{Z_R - Z}{Z + Z_s} - 1\right) \tag{6-48b}$$
$$= \frac{\bar{E}_a}{Z + Z_s}\left(\frac{Z_R - Z_s}{2} - Z\right)$$

By substituting Eq. (6-48) into Eq. (6-35), the torque generated by the magnitude comparator for phase A ground unit is found as

$$Tq = |\bar{S}_A|^2 - |\bar{S}_B|^2$$
$$= E_a^2 \frac{1}{|Z + Z_s|^2}\left(\left|\frac{Z_R + Z_s}{2}\right|^2 \right. \tag{6-49}$$
$$\left. - \left|Z - \frac{Z_R - Z_s}{2}\right|^2\right)$$

It can be verified that the torque described by Eq. (6-46) is the same as the torque described by Eq. (6-49).

3.2.2 Phase–Phase Units

All three kinds of phase-to-phase faults can be covered by a single polyphase unit. In the unit, the operating and polarizing signals are described as

$$\bar{S}_1 = \bar{V}_{ab} - Z_R \bar{I}_{ab} \tag{6-51a}$$
$$\bar{S}_2 = \bar{V}_{cb} - Z_R \bar{I}_{cb} \tag{6-51b}$$

Its trip condition is

$$0° \le \angle\left(\frac{\bar{S}_1}{\bar{S}_2}\right) \le 180°$$

This principle has been successfully applied to EM relays, solid-state relays, and microprocessor relays. Thus, we choose this unit to show the torque generated in phase–phase units. In the following, we demonstrate that torque generated in this unit is the same for different phase-to-phase faults: BC, CA, and AB faults.

BC Fault

For phase BC fault, the current in phase A is zero and the currents in phases B and C are equal in magnitude and 180° apart out-of-phase:

$$\bar{I}_a = 0 \tag{6-52a}$$

$$\bar{I}_b = \frac{\bar{E}_{bc}}{2(Z + Z_s)} \tag{6-52b}$$

$$\bar{I}_c = -\frac{\bar{E}_{bc}}{2(Z + Z_s)} \tag{6-52c}$$

At the relay location, the voltage on phase A is equal to the source voltage, the voltages on phases B and C are equal to source voltages minus the voltage drops by the corresponding currents:

$$\bar{V}_a = \bar{E}_a \tag{6-53a}$$
$$\bar{V}_b = \bar{E}_b - Z_s\bar{I}_b \tag{6-53b}$$
$$\bar{V}_c = \bar{E}_c - Z_s\bar{I}_c \tag{6-53c}$$

Substituting Eqs. (6-52) and (6-53) into Eq. (6-51), the operating and polarizing quantities of the phase comparator are described as

$$\bar{S}_1 = \bar{E}_{cb}\left[e^{-j\frac{\pi}{3}} - \frac{Z_R + Z_s}{2(Z + Z_s)}\right] \tag{6-54a}$$

$$\bar{S}_2 = \bar{E}_{cb}\left(1 - \frac{Z_R + Z_s}{Z + Z_s}\right) \tag{6-54b}$$

Substituting Eq. (6-54) into Eq. (6-34), the torque generated by the phase comparator for phase BC fault is found as

$$Tq = S_1 S_{2r} - S_1 S_{2i}$$
$$= \frac{\sqrt{3}}{2}E_{ab}^2 \frac{1}{|Z + Z_s|}[|Z_R + Z_s| \tag{6-55}$$
$$\cos(\phi_R' - \phi') - |Z + Z_s|]$$

CA Fault

For phase CA fault, the current in phase B is zero and the currents in phases C and A are equal in magnitude and 180° apart out-of-phase:

$$\bar{I}_a = \frac{\bar{E}_{ac}}{2(Z + Z_s)} \tag{6-56a}$$

$$\bar{I}_b = 0 \tag{6-56b}$$

$$\bar{I}_c = -\frac{\bar{E}_{ac}}{2(Z + Z_s)} \tag{6-56c}$$

At the relay location, the voltages on phases A, B, and C are described as

$$\bar{V}_a = \bar{E}_a - Z_s\bar{I}_a \tag{6-57a}$$

$$\bar{V}_b = \bar{E}_b \tag{6-57b}$$

$$\bar{V}_c = \bar{E}_c - Z_s\bar{I}_c \tag{6-57c}$$

If we substitute Eqs. (6-56) and (6-57) into Eq. (6-51), the operating and polarizing quantities of the phase comparator are described as

$$\bar{S}_1 = \bar{E}_{ac}\left[e^{j\frac{\pi}{3}} - \frac{Z_R + Z_s}{2(Z + Z_s)} \right] \tag{6-58a}$$

$$\bar{S}_2 = \bar{E}_{ac}\left[e^{j\frac{2\pi}{3}} + \frac{Z_R + Z_s}{2(Z + Z_s)} \right] \tag{6-58b}$$

By substituting Eq. (6-58) into Eq. (6-34), the torque generated by the phase comparator for phase CA fault is found as

$$
\begin{aligned}
Tq &= S_{1r}S_{2r} - S_{1r}S_{2i} \\
&= \frac{\sqrt{3}}{2}E_{ab}^2 \frac{1}{|Z + Z_s|}[|Z_R + Z_s| \\
&\quad \cos(\phi_R{}' - \phi') - |Z + Z_s|]
\end{aligned} \tag{6-59}
$$

AB Fault

For phase AB fault, the current in phase C is zero and the currents in phases A and B are equal in magnitude and 180° apart out-of-phase:

$$\bar{I}_a = \frac{\bar{E}_{ab}}{2(Z + Z_s)} \tag{6-60a}$$

$$\bar{I}_b = -\frac{\bar{E}_{ab}}{2(Z + Z_s)} \tag{6-60b}$$

$$\bar{I}_c = 0 \tag{6-60c}$$

At the relay location, the voltages on phases A, B, and C are described as

$$\bar{V}_a = \bar{E}_a - Z_s\bar{I}_a \tag{6-61a}$$

$$\bar{V}_b = \bar{E}_b - Z_s\bar{I}_b \tag{6-61b}$$

$$\bar{V}_c = \bar{E}_c \tag{6-61c}$$

Substituting Eqs. (6-60) and (6-61) into Eq. (6-51), the operating and polarizing quantities of the phase comparator are described as

$$\bar{S}_1 = \bar{E}_{ab}\left(1 - \frac{Z_R + Z_s}{Z + Z_s} \right) \tag{6-62a}$$

$$\bar{S}_2 = \bar{E}_{ab}\left[e^{j\frac{\pi}{3}} - \frac{Z_R + Z_s}{2(Z + Z_s)} \right] \tag{6-62b}$$

Substituting Eq. (6-62) into Eq. (6-34), the torque generated by the phase comparator for phase AB fault is found as

$$
\begin{aligned}
Tq &= S_{1r}S_{2r} - S_{1r}S_{2i} \\
&= \frac{\sqrt{3}}{2}E_{ab}^2 \frac{1}{|Z + Z_s|}[|Z_R + Z_s| \\
&\quad \cos(\phi_R{}' - \phi) - |Z + Z_s|]
\end{aligned} \tag{6-63}
$$

According to Eq. (6-46), (6-49), (6-50), and (6-55), the characteristic equation of ground and phase–phase distance units is described as

$$|Z - \rho| = \gamma \tag{6-64}$$

where

$$\gamma = \left| \frac{Z_R + Z_s}{2} \right|, \qquad \rho = \frac{Z_R - Z_s}{2}$$

Equation (6-64) represents the mho circle the center of which is at ρ and radius equal to γ as shown in Figure 6-27.

3.2.3 Torque Normalization

In observing Eqs. (6-46), (6-49), (6-50), and (6-55), we found that the torque created in ground distance units and phase–phase distance units has a similar expression. Dividing Eqs. (6-46) by E_a^2, (6-49) by E_a^2, and (6-55) by $(\sqrt{3}/2)E_a^2$, we obtain the normalized torque as

$$Tq^{norm} = \frac{1}{|Z_s + Z|^2}\left(\left| \frac{Z_R + Z_s}{2} \right|^2 - \left| Z - \frac{Z_R - Z_s}{2} \right|^2 \right) \tag{6-65}$$

Furthermore, if we assume the source impedance angle is the same as the line impedance angle, the impedance angles in Eq. (6-65) do not play any role in the final re-

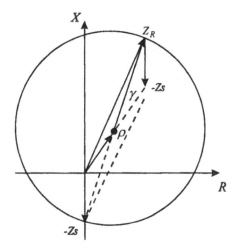

Figure 6-27 Mho circle of distance units.

sults. Thus, the normalized torque at the maximum torque angle can be expressed as

$$
\begin{aligned}
Tq^{norm} &= \frac{1}{(|Z_s| + |Z|)^2}\left[\left(\frac{|Z_R| + |Z_s|}{2}\right)^2 \right.\\
&\qquad \left. - \left(|Z| - \frac{|Z_R| - |Z_s|}{2}\right)^2\right]\\
&= \frac{|Z_R| - |Z|}{|Z_s| + |Z|} = \frac{1 - |Z|/|Z_R|}{SIR + |Z|/|Z_R|}
\end{aligned} \tag{6-66}
$$

where $SIR = |Z_s|/|Z_R|$ representing the source impedance ratio. Corresponding to different SIRs, the variations of the normalized torque relative to fault locations are il-

lustrated in Figure 6-28. It can be seen that the torque is negative beyond the reach boundary, zero at the reach boundary, and positive within the protected zone. The closer the fault is, the larger the torque will be. For a specific fault location, the torque is inversely proportional to the source impedance ratio.

3.3 Torque Computations

The previous section shows the relationships among the torque, fault impedance, and source impedance ratios. This section discusses methods for computing the torque based on sampled voltages and currents. OFPs are utilized to generate orthogonal current and voltage signals. In addition, the section provides an example of torque computations by using electromagnetic transient program (EMTP) simulated-data. In this example, both short-data window OFP and long-data window OFPs are applied and the torque results from different OFPs are compared.

Generally, the inputs to a phase comparator can be described in the discrete-time domain as

$$
s_1(k) = v_1(k) - k_{1r}i_1(k) - k_{1x}i_{1,or}(k) \tag{6-67a}
$$

$$
s_2(k) = k_2 v_2(k) - k_{2r}i_2(k) - k_{2x}i_{2,or}(k) \tag{6-67b}
$$

where k_{1r}, k_{1x}, k_2, k_{2r}, k_{2x} are constants, $v_1(k)$ and $v_2(k)$ are voltage samples, $i_1(k)$ and $i_2(k)$ are current samples, $i_{1,or}(k)$ is orthogonal to $i_1(k)$, and $i_{2,or}(k)$ is orthogonal to $i_2(k)$. By selecting appropriate currents, voltages, and constants, different relay units can be constructed. For example, the phase–phase unit discussed in Section 3.2.2 can be formulated by letting

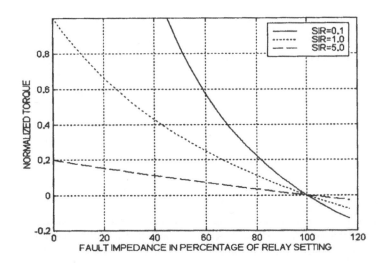

Figure 6-28 Normalized torque Tq^{norm} versus fault locations $|Z|/|Z_R|$.

$$v_1(k) = v_a(k) - v_b(k)$$
$$v_2(k) = v_c(k) - v_b(k)$$
$$i_1(k) = i_a(k) - i_b(k)$$
$$i_2(k) = i_c(k) - i_b(k)$$
$$k_{1r} = k_{2r} = R_R$$
$$k_{1x} = k_{2x} = X_R$$
$$k_2 = 1$$

where $v_a(k)$, $v_b(k)$, $v_c(k)$ represent voltage samples at instant k on phases A, B, and C, respectively, $i_a(k)$, $i_b(k)$, $i_c(k)$ represent current samples at instant k on phases A, B, and C, respectively, R_R and X_R represent relay resistance and reactance settings.

For the phase–phase unit, the relay trip condition is $s_1(k)$ leading $s_2(k)$:

$$0° \le \angle[s_1(k)] - \angle[s_2(k)] \le 180°$$

Thus, according to Eq. (6-34), the torque created in the phase comparator can be calculated as

$$Tq(k) = s_1(k)s_{2,or}(k) - s_{1,or}(k)s_2(k)$$

where $s_{1,or}(k)$ is orthogonal to $s_1(k)$; $s_{2,or}(k)$ is orthogonal to $s_2(k)$; and in addition,

$$s_{1,or}(k) = v_{1,or}(k) - k_{1r}i_{1,or}(k) - k_{1x}[i_{1,or}(k)]_{or} \quad (6\text{-}68a)$$
$$s_{2,or}(k) = k_2 v_{2,or}(k) - k_2 i_{2,or}(k) \quad\quad\quad (6\text{-}68b)$$
$$- k_{2x}[i_{2,or}(k)]_{or}$$

If the trip condition of a relay is the angle difference between $s_1(k)$ and $s_2(k)$ being less than 90°:

$$-90° \le \angle[s_1(k)] - \angle[s_2(k)] \le 90°$$

according to Eq. (6-32), the torque created in the corresponding phase comparator can be calculated as

$$Tq(k) = s_{1,or}(k)s_{2,or}(k) + s_1(k)s_2(k)$$

As shifting a signal 90° forward twice is equivalent to multiplying -1 to the signal, the last term in Eq. (6-68a) may be replaced by $-k_{1x}i_1(k)$, and the last term in Eq. (6-68b) may be replaced by $-k_{2x}i_2(k)$. Then, Eqs. (6-68a) and (6-68b) become

$$s_{1,or}(k) = v_{1,or}(k) - k_{1r}i_{1,or}(k) + k_{1x}i_1(k) \quad (6\text{-}69a)$$
$$s_{2,or}(k) = k_2 v_{2,or}(k) - k_2 i_{2,or}(k) + k_{2x}i_2(k) \quad (6\text{-}69b)$$

The torque created in a magnitude comparator can be computed in a similar way. For instance, if the inputs to a magnitude comparator are represented by $s_A(k)$ and $s_B(k)$, according to Eq. (6-35), the torque created in the magnitude comparator is computed as

$$Tq(k) = [s_A(k)^2 + s_{A,or}(k)^2] - [s_B(k)^2 + s_{B,or}(k)^2]$$

where $s_{A,or}(k)$ is orthogonal to $s_A(k)$, and $s_{B,or}(k)$ is orthogonal to $s_B(k)$.

The following is an example of torque computations based on the long-data window and short-data window OFPs. In this example, EMTP-simulated data is utilized. The system simulated contains two parallel lines with sources at both ends. The lines are 200 miles long and rated at 500 kV. The faults simulated are phase A to ground and the torque is computed in the phase A quadrature-polarized ground distance unit. The relay is set at 90% of the line impedance.

Figures (6-29)–(6-31) show the torque when the fault locations are at 20% (40 miles), 80% (160 miles), and 100% (200 miles) of the line, respectively. The time 0 in the figures represents the fault inception instant. The two-sample OFPs have an excellent dc-offset immunity, and the effect of the dc-offset on the associated torque is almost invisible. The half-cycle DFT OFPs respond poorly to the dc-offset, and there is a large overshoot in the torque. The torque overshoot is one of the reasons that cause a relay to overreach and lose security. The torque overshoot in the full-cycle DFT OFP is less severe than in the half-cycle DFT, but is much worse than in the two-sample OFP. Because of using short-data windows, the buildup torque speed of the two-sample OFP is faster, especially immediately following the fault inception. The two-sample OFPs are more sensitive to high-frequency noise, so the associated torque experiences small-magnitude high-frequency oscillations.

Observation of Figures (6-29)–(6-31) also shows the effect of fault locations on the torque. The torque is positive for faults within the protected zone and negative for faults outside the protected zone. The torque increases when the fault is closer to the relay location. Corresponding to faults at 20, 80, and 100% of the line, the steady-state torque is about 1.95, 0.08, and -0.14, respectively. The sign of the torque indicates whether a fault is within the protected zone, and the magnitude of the torque indicates how far a fault is from the reach boundary. The relay should be able to recognize both the sign and magnitude of the torque. The sign of the torque determines if the relay should trip, and the magnitude of the torque determines how fast the relay should trip to ensure speed and security.

3.4 Post-filters

As indicated earlier, the torque generated by a comparator can tell if a fault is within the protected zone as well as how far the fault is from the relay location. To ensure a fast trip for close-in faults and a slower but secure trip for near boundary faults, post-filters which process torque (raw trip signal) are used and are shown in Figure 6-32. This section

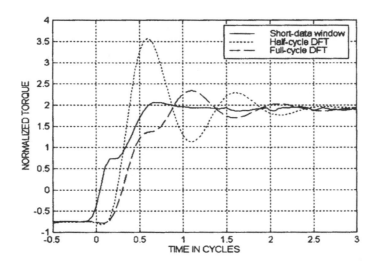

Figure 6-29 Phase A-to-ground fault at 20% of the line with relay setting at 90%.

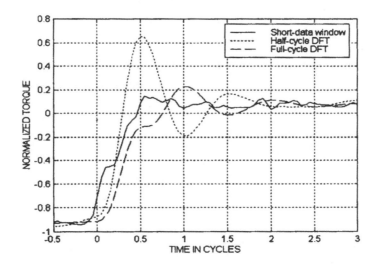

Figure 6-30 Phase A-to-ground fault at 80% of the line, with relay setting at 90%.

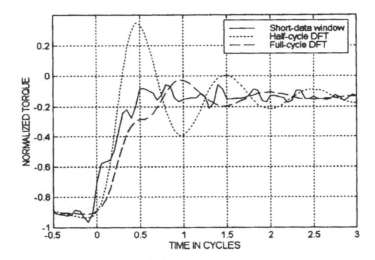

Figure 6-31 Phase A-to-ground fault at 100% of the line, with relay setting at 90%.

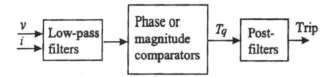

Figure 6-32 Post-filters utilized in a relay unit.

presents two kinds of post-filters. One post-filter is based on the emulator of the dynamics of the EM cylinder unit, and the other is based on an adaptive trip counter.

3.4.1 Emulator of the Dynamics of an EM Cylinder Unit

In an EM distance relay using a cylinder unit, the torque generated in the unit turns a cylinder and thereby moves the relay contact toward a trip direction if a fault is within the protected zone. The rotating angle θ of the cylinder is related to the torque by the following dynamic equation

$$Tq = M\frac{d^2\theta}{dt^2} + K_d\frac{d\theta}{dt} + K_c\theta \tag{6-70}$$

where M, K_d, and K_c are the cylinder inertia constant, damping constant, and spring constant, respectively. By selecting appropriate constants, various kinds of inverse characteristics may be obtained.

3.4.2 Adaptive Trip Counter

An adaptive trip counter can also be applied to accomplish the desirable inverse characteristics. The trip counter counts the samples of positive torque and initiates trip signals based on both the counted sample number and torque magnitudes. Assume that the relay is allowed to trip fast if the normalized torque is larger than a constant ϵ, then

$$\frac{1}{|Z_s + Z|^2}(\gamma^2 - |Z - \rho|^2) = \epsilon \tag{6-71}$$

By using their real and imaginary parts to represent complex quantities, Eq. (6-71) can be rearranged as

$$(R - \rho_r)^2 + (X - \rho_i)^2 \tag{6-72}$$
$$+ \epsilon[(R + R_s)^2 + (X + X_s)^2] = \gamma^2$$

or

$$R^2 + X^2 - 2\frac{\rho_r - \epsilon R_s}{1 + \epsilon}R - 2\frac{\rho_i - \epsilon X_s}{1 + \epsilon}X \tag{6-73}$$

$$= \frac{1}{1 + \epsilon}[\gamma^2 - \rho^2 - \epsilon(R_s^2 + X_s^2)]$$

Obviously, Eq. (6-73) represents a circle the radius of which is

$$\gamma' = \frac{1}{1 + \epsilon}\gamma$$

and center at

$$\left(\frac{\rho_r - \epsilon R_s}{1 + \epsilon}, \frac{\rho_i - \epsilon X_s}{1 + \epsilon}\right)$$

Figure 6-33 shows the mho circle of the fast-operating region, represented by the dotted circle, along with the original mho circle. It can be seen that both circles pass the point $-Z_s$ and their centers are on the same line that starts from $-Z_s$ and ends at Z_c.

4. EXPLICIT (IMPEDANCE) RELAY ALGORITHMS

A direct way in designing a distance relay unit is to compute the faulted line impedance, based on sampled voltage and current, and then compare it with the impedance setting—the explicit method. For different types of faults, suitable voltage and current may be selected such that they satisfy

$$v(k) = Ri(k) + Xi_{or}(k) \tag{6-74}$$

where R and X represent the line-positive sequence resistance and reactance from the relay to the fault location.

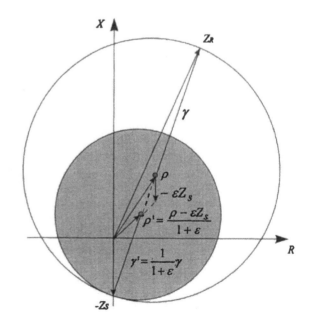

Figure 6-33 Mho circle of fast-operating region (shaded area).

Equation (6-74) has two unknowns: R and X. Thus, two equations are required to solve for them. After R and X are found, they are compared with the relay settings R_R and X_R to determine if a fault is within the protected zone. This section presents two methods for formulating the two necessary equations to solve for R and X.

Method 1   Use $v(k)$ and $v_{or}(k)$—orthogonal signal pair.
Based on $v(k)$ and $v_{or}(k)$, the simultaneous equations for R and X can be expressed as

$$v(k) = Ri(k) + Xi_{or}(k) \tag{6-75a}$$

$$v_{or}(k) = Ri_{or}(k) - Xi(k) \tag{6-75b}$$

In obtaining Eq. (6-75b), we take advantage of the fact that shifting the phase angle of a signal 90° ahead twice is equivalent to multiplying the signal by -1. Solving Eqs. (6-75a) and (6-75b), the line positive-sequence resistance R and reactance X are found as

$$R = \frac{i(k)v(k) + i_{or}(k)v_{or}(k)}{i(k)^2 + i_{or}(k)^2} \tag{6-76a}$$

$$X = \frac{i_{or}(k)v(k) - i(k)v_{or}(k)}{i(k)^2 + i_{or}(k)^2} \tag{6-76b}$$

Method 2   Use $v(k)$ and $v(k-n)$—voltages at different sampling instants.
Based on $v(k)$ and $v(k-n)$, the simultaneous equations for R and X can be expressed as

$$v(k) = Ri(k) + Xi_{or}(k) \tag{6-77a}$$

$$v(k-n) = Ri(k-n) + Xi_{or}(k-n) \tag{6-77b}$$

where n is the time delay in samples. Solving Eqs. (6-77a) and (6-77b), the resistance and reactance are found as

$$R = \frac{i_{or}(k-n)v(k) - i_{or}(k)v(k-n)}{i(k)i_{or}(k-n) - i_{or}(k)i(k-n)} \tag{6-78a}$$

$$X = \frac{i(k)v(k-n) - i(k-n)v(k)}{i(k)i_{or}(k-n) - i_{or}(k)i(k-n)} \tag{6-78b}$$

Similar to the situation in the implicit (torque-like) relay algorithms, the critical task in implementing the ex-

plicit relay algorithms is to obtain orthogonal signal pairs. Various OFPs described in previous sections may be utilized to create the orthogonal signals. If long-data window OFPs are used, the response speed of R and X will be slower, but they will be less affected by high-frequency noise. On the other hand, if short-data window OFPs are used, the response speed of R and X will be faster, but the solutions may experience high-frequency oscillations. As in the implicit algorithms, post-filters need to be used in explicit algorithms to obtain the desirable inverse characteristics.

In the following examples, short-data window OFPs are used in generating the necessary orthogonal signals. The short-data window OFP utilized is algorithm 2 of the two-sample OFPs. The input–output relationship in the two-sample OFP is given in Eqs. (6-10a) and (6-10b) and repeated here for easy reference

$$y(k) = \frac{1}{2\cos(\omega_0 T/2)}[x(k) + x(k-1)]$$

$$y_{or}(k) = \frac{1}{2\sin(\omega_0 T/2)}[x(k) - x(k-1)]$$

Therefore, the orthogonal signal pair derived form the current samples are

$$\frac{1}{2\cos(\omega_0 T/2)}[i(k) + i(k-1)] \tag{6-79a}$$

$$\frac{1}{2\sin(\omega_0 T/2)}[i(k) - i(k-1)] \tag{6-79b}$$

and the orthogonal signal pair derived form the voltage samples are

$$\frac{1}{2\cos(\omega_0 T/2)}[v(k) + v(k-1)] \tag{6-80a}$$

$$\frac{1}{2\sin(\omega_0 T/2)}[v(k) - v(k-1)] \tag{6-80b}$$

For method 1, substituting Eqs. (6-79a) and (6-79b) into Eqs. (6-76a) and (6-76b) for $i(k)$ and $i_{or}(k)$, respectively, and substituting Eqs. (6-80a) and (6-80b) into Eqs. (6-76a) and (6-76b) for $v(k)$ and $v_{or}(k)$, respectively, R and X are found as (see below).

$$R = \frac{\tan^2(\omega_0 T/2)[i(k) + i(k-1)][v(k) + v(k-1)] + [i(k) - i(k-1)][v(k) - v(k-1)]}{\tan^2(\omega_0 T/2)[i(k) + i(k-1)]^2 + [i(k) - i(k-1)]^2}$$

$$X = \frac{2[i(k)v(k-1) - i(k-1)v(k)]}{\tan(\omega_0 T/2)[i(k) + i(k-1)]^2 + \text{ctan}(\omega_0 T/2)[i(k) - i(k-1)]^2}$$

Method 1

$$R = \frac{[i(k-1) - i(k-2)][v(k) + v(k-1)] - [i(k) - i(k-1)][v(k-1) + v(k-2)]}{[i(k) + i(k-1)][i(k-1) - i(k-2)] - [i(k) - i(k-1)][i(k-1) + i(k-2)]}$$

$$X = \tan\left(\frac{\omega_0 T}{2}\right)\frac{[i(k) + i(k-1)][v(k-1) + v(k-2)] - [i(k-1) + i(k-2)][v(k) + v(k-1)]}{[i(k) + i(k-1)][i(k-1) - i(k-2)] - [i(k) - i(k-1)][i(k-1) + i(k-2)]}$$

Method 2

For method 2, substituting Eqs. (6-79a) and (6-79b) into Eqs. (6-78a) and (6-78b) for $i(k)$ and $i_{or}(k)$, respectively, and substituting Eqs. (6-80a) and (6-80b) into Eqs. (6-78a) and (6-78b) for $v(k)$ and $v_{or}(k)$, respectively, R and X are found as (see above).

5. SYMMETRICAL COMPONENT FILTERS

Symmetrical component method is a very powerful tool in fault analysis and protective relay designs. Utilization of symmetrical components allows engineers to have a wider view and thus make more intelligent choices in selecting electric quantities in the designs of protective relays. Typical relay units that utilize symmetrical components include zero-sequence–polarized directional units, negative-sequence–polarized directional units, negative- and zero-sequence currents-based fault-phase selectors, and combined-sequence current differential and phase comparison relays.

According to symmetrical components method, any unbalanced three-phase currents or voltages in a power system can be represented by three balanced sets of currents or voltages. The three sets are called positive-, negative-, and zero-sequence. In the following, currents are used to illustrate the symmetrical components concept. Based on the effects of phase currents for phases A, B, and C, the

positive-, negative-, and zero-sequence currents on phase A are determined by

$$I_{a1} = \frac{1}{3}(I_a + aI_b + a^2I_c) \tag{6-81a}$$

$$I_{a2} = \frac{1}{3}(I_a + a^2I_b + aI_c) \tag{6-81b}$$

$$I_{a0} = \frac{1}{3}(I_a + I_b + I_c) \tag{6-81c}$$

where I_a, I_b, and I_c represent currents on phases A, B, and C; I_{a1}, I_{a2}, and I_{a0} represent positive-, negative-, and zero-sequence currents on phase A; a and a^2 are phase shifters and

$$a = e^{j120°} = -\frac{1}{2} + j\frac{\sqrt{3}}{2} \tag{6-82}$$

$$a^2 = e^{-j120°} = -\frac{1}{2} - j\frac{\sqrt{3}}{2} \tag{6-83}$$

Sequence quantities on phase B and phase C are related to phase A sequence quantities by

$$\begin{aligned} I_{b1} &= a^2I_{a1} & I_{c1} &= aI_{a1} \\ I_{b2} &= aI_{a1} & I_{c2} &= a^2I_{a2} \\ I_{b0} &= I_{c0} = I_{a0} \end{aligned}$$

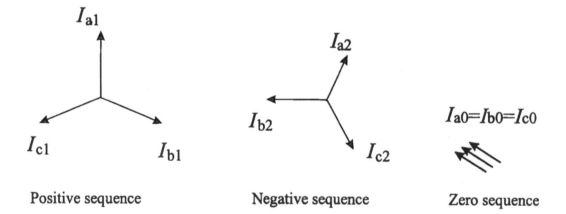

Positive sequence   Negative sequence   Zero sequence

Figure 6-34 Relation between sequence quantities in different phases.

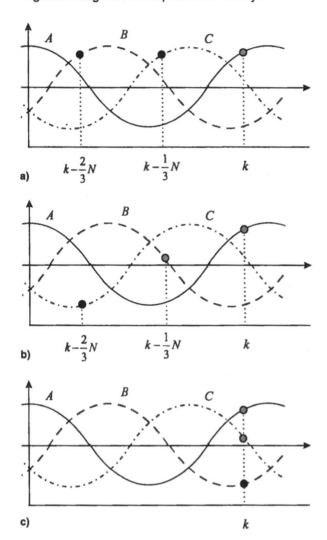

a)

$$k - \frac{2}{3}N \qquad k - \frac{1}{3}N \qquad k$$

b)

$$k - \frac{2}{3}N \qquad k - \frac{1}{3}N \qquad k$$

c)

$$k$$

Figure 6-35 Symmetrical components by time-delaying: (a) positive-sequence quality at k; (b) negative-sequence quantity at k; (c) zero sequence quantity at k.

The relation between sequence quantities in phases A, B, and C may well be illustrated by phasor diagram as shown in Figure 6-34. The positive-sequence current I_{a1} leads I_{b1} which, in turn, leads I_{c1} by 120°; the negative-sequence current I_{a2} lags I_{b2} which, in turn, lags I_{c2} by 120°; and the zero-sequence currents I_{a0}, I_{b0}, and I_{c0} are in-phase.

On each phase, the phase quantity is the sum of its sequence quantities, that is,

$$I_a = I_{a1} + I_{a2} + I_{a0}$$
$$I_b = I_{b1} + I_{b2} + I_{b0}$$
$$I_c = I_{c1} + I_{c2} + I_{c0}$$

Among positive-, negative-, and zero-sequence quantities, the zero-sequence quantity is the easiest to obtain because it takes a direct sum of phases quantities on phases A, B, and C. But for positive- and negative-sequence quantities, either a 120° or −120° shift needs to be accomplished.

Various approaches have been proposed to extract symmetrical components from phase quantities. This section focuses only on numerical methods and presents three different approaches for computing symmetrical components.

5.1 Phase Shift by Time-Delaying

One method to obtain a phase shift is through time-delaying. Because no extra computations are introduced in such a process, this method is easier to implement. For example, the sample taken two-thirds cycles earlier may be used to obtain a 120° (or −240°) phase shift, and similarly, the sample taken one-third cycles earlier may be used to obtain a −120° phase shift from the most recent sample. Therefore, by time-delaying, the positive- and negative-sequence currents on phase A are obtained as:

$$i_{a1}(k) = \frac{1}{3}\left[i_a(k) + i_b\left(k - \frac{2}{3}N\right) + i_c\left(k - \frac{1}{3}N\right) \right] \quad (6\text{-}84)$$

$$i_{a2}(k) = \frac{1}{3}\left[i_a(k) + i_b\left(k - \frac{1}{3}N\right) + i_c\left(k - \frac{2}{3}N\right) \right] \quad (6\text{-}85)$$

where N represents the sampling rate, $-2N/3$ represents a two-thirds cycle time-delaying, and $-N/3$ represents a one-third cycle time-delaying. Figure 6-35 shows the process of extracting positive-, negative-, and zero-sequence components by using time-delaying method.

Although the time-delaying method is simple, it is slower and requires two-thirds cycles to complete the computations. Thus, the method is more suitable for applications where relay operating speed is not critical.

5.2 Phase Shift by Orthogonal Signals

In addition to the time-delaying technique, the ±120° phase shifts may also be accomplished by using orthogonal signals generated by OFPs. For this purpose, substituting the polar form of the phase shifters a and a^2 into Eqs. (6-81a) and (6-81b), we have the positive- and negative-sequence currents:

$$I_{a1} = \frac{1}{6}[2I_a - I_b - I_c + j\sqrt{3}(I_b - I_c)] \quad (6\text{-}86a)$$

$$I_{a2} = \frac{1}{6}[2I_a - I_b - I_c + j\sqrt{3}(I_b - I_c)] \quad (6\text{-}86b)$$

$$I_{a0} = \frac{1}{3}(I_a + I_b + I_c) \quad (6\text{-}86c)$$

It can be shown that, in terms of the Clark's components method, which is another helpful tool for fault analysis, Eqs. (6-86a) and (6-86b) can be reformulated as:

$$I_{a1} = \frac{1}{2}(I_\alpha + jI_\beta) \qquad (6\text{-}87a)$$

$$I_{a2} = \frac{1}{2}(I_\alpha - jI_\beta) \qquad (6\text{-}87b)$$

where I_α and I_β are Clark α and β components:

$$I_\alpha = \frac{1}{3}(2I_a - I_b - I_c)$$

$$I_\beta = \frac{1}{\sqrt{3}}(I_b - I_c)$$

Obviously, calculations of the α and β components are straightfoward and no phase shift is required:

$$i_\alpha(k) = \frac{1}{3}[2i_a(k) - i_b(k) - i_c(k)]$$

$$i_\beta(k) = \frac{1}{\sqrt{3}}[i_b(k) - i_c(k)]$$

Once α and β components are obtained, the positive- and negative-sequence components can be computed by using orthogonal signals provided by the OFPs. In the following, we show two examples for obtaining symmetrical components based on OFPs.

The first example is based on algorithm 1 of the two-sample OFPs. The input–output relationship in the two-sample OFPs is

$$y(k) = x(k)$$

$$y_{or}(k) = \frac{1}{\sin \omega_0 T}[\cos \omega_0 T x(k) - x(k-1)]$$

where $y_{or}(k)$ leads $y(k)$ by 90°. Substituting these two equations into Eqs. (6-87a) and (6-87b), we get the symmetrical components:

$$i_{a1}(k) = \frac{1}{2}\left\{i_\alpha(k) + \frac{1}{\sin \omega_0 T}[\cos \omega_0 T i_\beta(k) - i_\beta(k-1)]\right\}$$

$$i_{a2}(k) = \frac{1}{2}\left\{i_\alpha(k) + \frac{1}{\sin \omega_0 T}[\cos \omega_0 T i_\beta(k) - i_\beta(k-1)]\right\}$$

The corresponding zero sequence current is

$$i_{a0}(k) = \frac{1}{3}[i_a(k) + i_b(k) + i_c(k)]$$

The second example is based on quarter-cycle time-delaying OFPs. The input–output relationship in the quarter-cycle time-delaying OFPs is

$$y(k) = x\left(k - \frac{N}{4}\right)$$

$$y_{or}(k) = x(k)$$

In the OFPs, the 90° phase shift is accomplished by a quarter-cycle time-delaying. Substituting these two equations into Eqs. (6-87a) and (6-87b), we obtain the symmetrical components:

$$i_{a1}(k) = \frac{1}{2}\left[i_\alpha\left(k - \frac{N}{4}\right) + i_\beta(k)\right]$$

$$i_{a2}(k) = \frac{1}{2}\left[i_\alpha\left(k - \frac{N}{4}\right) - i_\beta(k)\right]$$

The corresponding zero sequence current is

$$i_{a0}(k) = \frac{1}{3}\left[i_a\left(k - \frac{N}{4}\right) + i_b\left(k - \frac{N}{4}\right) + i_c\left(k - \frac{N}{4}\right)\right]$$

5.3 Two-Sample Phase Shifter

In the discrete time domain, two consecutive samples have a phase disagreement equal to 2π divided by the sampling rate N. To obtain a phase shift other than $2\pi/N$ by using two consecutive samples, the two samples need to be manipulated arithmetically. As a matter of fact, the weighted difference between two consecutive samples can produce any phase shift in range from zero to 2π. Such a two-sample phase shifter may be expressed as

$$y(k) = a_0 x(k) - a_1 x(k-1) \qquad (6\text{-}88)$$

where a_0 and a_1 are constants, $x(k)$ and $y(k)$ represent the input and output of the two-sample phase shifter. In phasor form, Eq. (6-88) can be described as

$$Y = (a_0 - a_1 e^{-j\omega_0 T})X \qquad (6\text{-}89)$$

The phasor relationship between X and Y is illustrated in Figure 6-36.

From Figure 6-36 it is found, intuitively, that any phase shift between X and Y can be obtained by selecting appropriate weighting constants a_0 and a_1. Let γ represent the desired phase shift between the input and output, then, the weighting constants a_0 and a_1 have to satisfy the following equation:

$$a_0 - a_1 e^{-j\omega_0 T} = e^{j\gamma} \qquad (6\text{-}90)$$

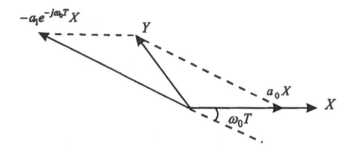

Figure 6-36   Phasor diagram of the two-sample phase shifter.

By separating Eq. (6-90) into real and imaginary parts, we have

$$a_0 - a_1 \cos \omega_0 T = \cos \gamma \tag{6-91}$$
$$a_1 \sin \omega_0 T = \sin \gamma \tag{6-92}$$

The constants a_0 and a_1 are obtained by solving the simultaneous Eqs. (6-91) and (6-92):

$$a_0 = \cos \gamma + \frac{\sin \gamma}{\tan \omega_0 T} \tag{6-93}$$

$$a_1 = \frac{\sin \gamma}{\sin \omega_0 T} \tag{6-94}$$

Algorithm 1 of the two-sample OFPs is, in fact, a specific application of the two-sample phase shifter where the phase shift is equal to 90°.

In computing symmetrical components, we need both 120° (a) and −120° (a^2) phase shifters. To obtain a 120° phase shift, let γ in Eqs. (6-93) and (6-94) equal 120°, then the constants are found as

$$a_0 = -\frac{1}{2} + \frac{\sqrt{3}}{2} \operatorname{ctan} \omega_0 T \tag{6-95}$$

$$a_1 = \frac{\sqrt{3}}{2} \frac{1}{\sin \omega_0 T} \tag{6-96}$$

Similarly, to obtain a −120° phase shift, let γ equal −120°, then the constants are

$$a_0 = -\frac{1}{2} - \frac{\sqrt{3}}{2} \operatorname{ctan} \omega_0 T \tag{6-97}$$

$$a_1 = -\frac{\sqrt{3}}{2} \frac{1}{\sin \omega_0 T} \tag{6-98}$$

Based on phase shifter constants given in Eqs. (6-95)–(6-98), the positive- and negative-sequence currents are computed as

$$i_{a1}(k) = \frac{1}{3}[i_a(k)$$

$$-\frac{1}{2}(1 - \sqrt{3} \operatorname{ctan} \omega_0 T)i_b(k) \tag{6-99}$$

$$-\frac{\sqrt{3}}{2}\frac{1}{\sin \omega_0 T}i_b(k-1)$$

$$-\frac{1}{2}(1 + \sqrt{3} \operatorname{ctan} \omega_0 T)i_c(k) + \frac{\sqrt{3}}{2}\frac{1}{\sin \omega_0 T}i_c(k-1)$$

$$i_{a2}(k) = \frac{1}{3}[i_a(k)$$

$$-\frac{1}{2}(1 + \sqrt{3} \operatorname{ctan} \omega_0 T)i_b(k) \tag{6-100}$$

$$+\frac{\sqrt{3}}{2}\frac{1}{\sin \omega_0 T}i_b(k-1)$$

$$-\frac{1}{2}(1 - \sqrt{3} \operatorname{ctan} \omega_0 T)i_c(k) - \frac{\sqrt{3}}{2}\frac{1}{\sin \omega_0 T}i_c(k-1)]$$

The corresponding zero sequence current is

$$i_{a0}(k) = \frac{1}{3}[i_a(k) + i_b(k) + i_c(k)]$$

As an example, let us assume the sampling rate to be 12 samples per cycle and find the coefficients in the two-sample ±120° phase shifters. For a 12 samples per cycle sampling rate, the phase delay between two consecutive samples is

$$\omega_0 T = 30°$$

Substituting $\omega_0 T$ into Eqs. (6-95)–(6-98), we obtain the constants for the 120° phase shifter:

$$a_0 = 1, \qquad a_1 = \sqrt{3}$$

and for the $-120°$ phase shifter:

$$a_0 = -2, \qquad a_1 = -\sqrt{3}$$

Substituting the phase shifter constants into Eqs. (6-81a) and (6-81b), the positive- and negative-sequence components are

$$i_{a1}(k) = \frac{1}{3}[i_a(k) + i_b(k) - \sqrt{3}i_b(k-1)$$
$$- 2i_c(k) + \sqrt{3}i_c(k-1)]$$

$$i_{a2}(k) = \frac{1}{3}[i_a(k) - 2i_b(k) + \sqrt{3}i_b(k-1)$$
$$+ i_c(k) - \sqrt{3}i_c(k-1)]$$

6. SUMMARY

This chapter has presented two microprocessor-based distance relaying algorithms: implicit (torque-like) algorithms and explicit (impedance) algorithms. In implicit algorithms, fault distances are implicitly compared with distance settings by the means of torque, which may be generated by phase comparators or magnitude comparators. In explicit algorithms, fault distances are computed in terms of fault impedance and then explicitly compared with distance settings. In both implicit and explicit relaying algorithms, the desirable inverse characteristics may be obtained by using post-filters, that allow a relay to trip extremely faster for close-in faults and slower for reach boundary faults.

The essential task in implementing both implicit and explicit relaying algorithms is obtaining orthogonal signals. For this purpose, both short-data window and long-data window orthogonal filter pairs (OFPs) have been presented. The short-data window OFPs utilize the most recent two or three samples to obtain orthogonal signals, whereas the long-data window OFPs utilize the most recent one or half cycle's data to obtain orthogonal signals. Compared with long data-window OFPs, short-data window OFPs do not need long memory and respond fast to faults. But short-data window OFPs cannot attenuate high-frequency noise, so they have to be used in combinations with other filters.

This chapter has also presented numerical methods for extracting symmetrical components from phase quantities. The fundamental algorithms for determining symmetrical components are obtaining $120°$ and $240°$ phase shifts. Three phase-shift methods are discussed: time-delaying, orthogonal signals, and two-sample phase shifter. The time-delaying method takes advantage of the fact that two samples taken one third cycle apart in time have a phase displacement of $120°$, and two samples taken two thirds cycle apart in time have a phase displacement of $240°$. In the orthogonal signal method, $120°$ and $240°$ phase shifts are accomplished by using two orthogonal signals. Finally, in the two-sample method, two consecutive samples are utilized to obtain the required $120°$ and $240°$ phase shifts.

7

Series-Compensated Line Protection Philosophies

WALTER A. ELMORE

1. INTRODUCTION

No relaying application presents a greater challenge than the high-speed relaying of transmission lines equipped with series capacitors. This chapter describes many of the pitfalls that exist in applying conventional relaying practices and will describe the extremes to which we have gone to circumvent these problems. The chapter will describe several types of relaying systems that have been designed with the series capacitor application in mind. Much of the information in this chapter was provided by Finn Andersson, formerly of ABB Vasteras, Sweden, whose insightful contributions were of great value.

2. BACKGROUND

Transmission lines are inherently inductive. The purpose of a series capacitor is to tune out part or all of this inductance. In a network without series capacitors, faults are inductive in character, and the current will always lag the voltage by some angle. Commonly used types of line protection can detect the fault, and by operating circuit breakers clear it fast and selectively. With the series compensation of the transmission line, capacitive elements are introduced, and the network will no longer be inductive

under all fault conditions. The degree of this change is dependent on the line and network parameters, the extent of series compensation, the type of fault, and the fault location.

The capacitive, or apparent capacitive, nature of the fault current may cause failure of the line protection to operate, or to operate incorrectly, unless careful measures are taken to acknowledge this problem. Owing to the capacitive nature of the fault loop, the complication relative to the protection may arise, on both the compensated line as well as the adjacent lines.

Series capacitor banks are equipped with spark-gaps that protect the capacitor, and often with metal oxide protective devices. The spark-gaps are set to flash over at a voltage two to three times the nominal voltage of the bank. When the spark-gaps flash over, the network is restored to an inductive nature. In spite of this, the protection complications remain. The spark-gaps do not flash instantaneously following the occurrence of a fault, and do not flash at all under some fault conditions. The time to gap-flashing is often longer than the operating time of high-speed line protection. The effect of capacitive reactances must be evaluated even for faults that flash the spark-gaps. Adding to these complications is that severe transients are generated because of the presence of the series capacitor at the occurrence of the fault as well as at the instant of spark-gap flashing.

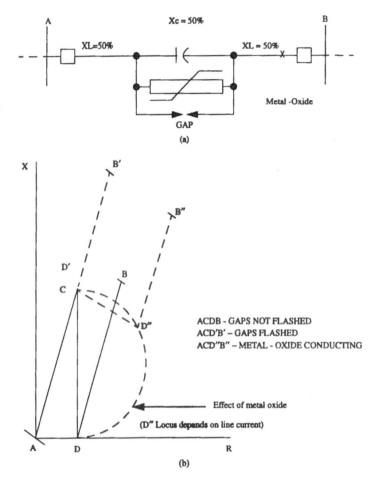

Figure 7-1 Apparent impedance as viewed from station "A" for a fault at B.

3. RELAYING QUANTITIES UNDER FAULT CONDITIONS

3.1 Series Compensation

The effect of series compensation on transmission line protection depends on the location of the capacitors and the degree of compensation. Figure 7-1a shows an example of a one-line diagram of a series capacitor and a transmission line. Figure 7-1b is the "steady-state" R–X (resistance–reactance) diagram. Because the capacitor-bypass protective equipment may be conducting or not, the apparent impedance as viewed from location A for a fault at B may appear vastly different.

3.2 Negative Reactance Effect

Figure 7-2 shows the influence of a nearby series capacitor bank. Faults near the capacitor-line junction as viewed from location A will have a very large negative-reactance

The use of metal oxide devices, in parallel with the series capacitor, introduces another element of concern. These units are never removed unless they themselves are jeopardized. Their level of conduction is approximately two times the rated peak voltage of the series capacitor. When voltage in excess of their conduction level appears across the metal oxide device, their impedance reduces markedly, causing the series capacitor to be partly bypassed. However, when the voltage decreases to a level below the threshold, the impedance of the device becomes very high, and the capacitor is effectively reinserted. This action provides another level of transient generation, but in general, causes a softer impact on the protective relaying than simple spark-gaps.

The metal oxide devices are bridged with triggered spark-gaps to limit the energy generated in the device during fault conditions. Therefore, the protective relays must be able to handle the effect of both the metal oxide device and spark-gaps.

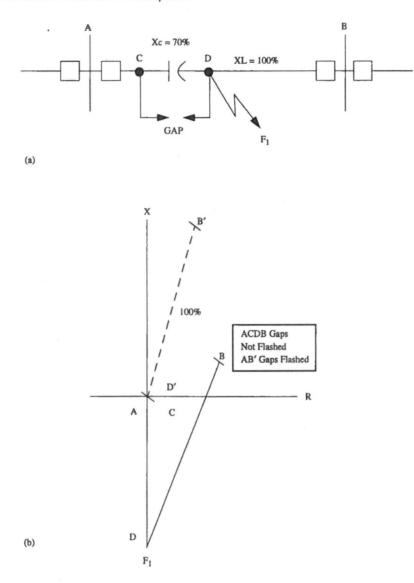

Figure 7-2   Apparent impedance at 60-Hz under fault conditions.

character. This negative reactance is actually due to reversal of the voltage at the relaying point or under certain conditions reversal of the current through the series capacitor.

3.3   Voltage Reversal

Voltage reversal occurs at the bus if the negative reactance of the series capacitor is greater than the positive reactance of the line section to the fault location. Current reversal occurs if the negative reactance of the series capacitor is greater than the sum of the source reactance and the line reactance to the fault location. Figure 7-3 shows this condition.

3.4   Zero Voltage Point

As can be seen in Figure 7-3a the "zero" voltage point in the system can be moved farther back in the system as a result of multiple lines contributing to a fault near the capacitor-line junction. The negative reactance of the capacitor is enlarged compared with the positive reactance of one of the adjacent lines. This can result in zero voltage occurring on lines that are located far away from the series capacitor. The voltage can be zero only in a network with negligible resistances. In a real network the remaining voltage is so small that it can be considered zero.

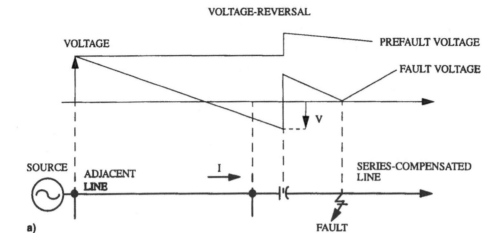

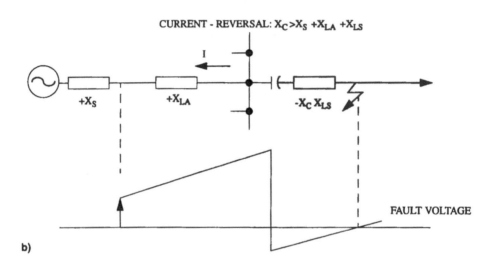

Figure 7-3 Voltage and current reversal.

3.5   Current Direction

"Current reversal" can also occur in some applications where a fault is at the capacitor line junction and a parallel line exists between buses A and B. Figure 7-4 describes the variation of voltage to be expected, for a typical case, at various locations in the power system. Note that there is no voltage inversion here. "Current inversion" occurs at 2; current at 4, is also in a direction opposite that for the same case without series capacitors. Whether line-side or bus potentials are used for the relays makes no difference in establishing the direction to this fault. However, distance relays are very sensitive to the location of the potential supply. Figures 7.4b–i show the difference in apparent impedance that distance relays would observe for this one fault.

For this example the distance relays at 1 and 3 operate correctly: 1 trips and 3 restrains. The relay at 2 will restrain undesirably or trip on memory action, depending on where the relay potential is obtained. The relay at 4 has the wrong directional sense and may trip undesirably if the distance unit is set to reach far enough. This is unacceptable and special logic must be included to avoid this incorrect identification of fault location.

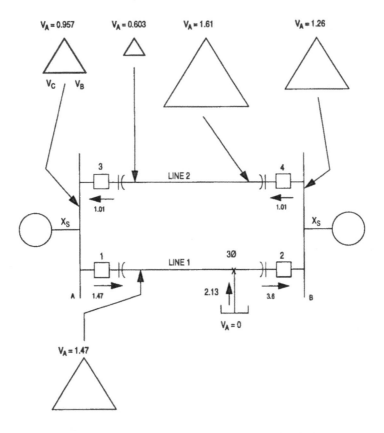

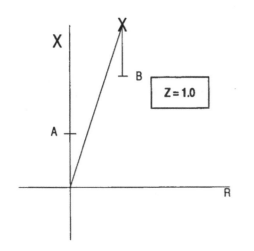

Figure 7-4a   Typical voltages and currents for fault at capacitor-line junction ($X_S = 0.1$; $X_C = -0.35$, $X_L = 1.0$).

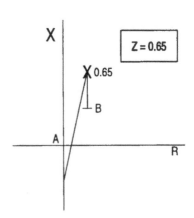

Figure 7-4b   Relay 1 line-side Vt's.

Figure 7-4c   Relay 1 bus-side Vt's.

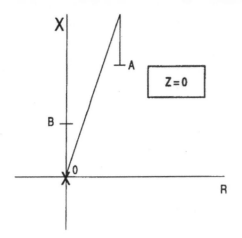

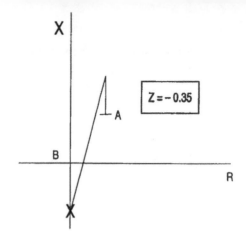

Figure 7-4d Relay 2 line-side Vt's.

Figure 7-4e Relay 2 bus-side Vt's.

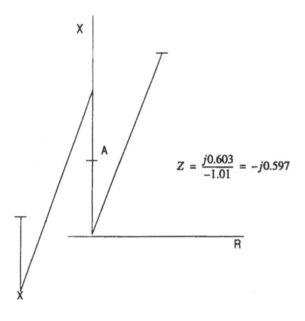

$$Z = \frac{j0.603}{-1.01} = -j0.597$$

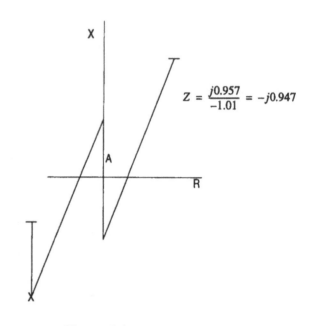

$$Z = \frac{j0.957}{-1.01} = -j0.947$$

Figure 7-4f Relay 3 line-side Vt's.

Figure 7-4g Relay 3 bus-side Vt's.

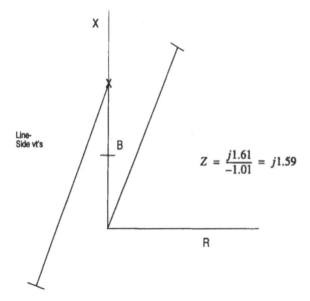

Line-Side vt's

$$Z = \frac{j1.61}{-1.01} = j1.59$$

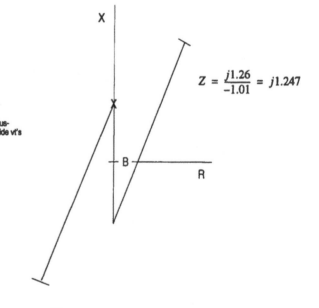

Bus-Side vt's

$$Z = \frac{j1.26}{-1.01} = j1.247$$

Figure 7-4h Relay 4 line-side Vt's.

Figure 7-4i Relay 4 bus-side Vt's.

3.6 Speed of Transition

When a fault occurs on an uncompensated transmission line, the impedance presented to the relays swings instantaneously from that representing the load to that representing the fault condition. Because of a decaying low-frequency transient component present in faults involving series-compensated lines, progress from the load point to the fault point on fault incidence may not be instantaneous. Instead, it may follow a logarithmic spiral, as indicated in Figure 7-5. Large amounts of compensation can result in the fault taking as long as 100 ms to progress from the load to the fault point.

4. DISTANCE PROTECTION

Series compensation of a network will affect the distance protection on both the compensated line and adjacent lines connected to bus bars at which a voltage reversal can occur. Generally, the most severe problems occur with the relaying associated with the adjacent line.

The following problem areas can be identified:

- Determination of direction to a fault
- Low frequency oscillation
- Transient caused by flashing of bridging gaps
- Transfer of capacitive reactance to resistance by metal oxide element bridging the capacitor
- Zone reach measurement
- False voltage "zeros" on adjacent healthy lines

Distance protection will experience difficulties in determining the correct direction to a fault in a station where a voltage reversal can occur. If direct polarization (polarization voltage from the faulty phase) is used, the protection on both the faulty and healthy lines may see the fault to be in the improper direction. This false direction determination will take place with both mho relays and plain directional elements.

To overcome this and achieve correct directional measurement, polarization quantities from the healthy phases are utilized. Healthy phase quantities will not be reversed, and a correct directional measurement will be achieved for all unsymmetrical faults for an unlimited time.

Cross-polarized mho relays will, under some fault conditions on adjacent lines overtrip because of the use of a single comparator for both the direction and reach measurement, and therefore additional measuring criteria are required.

For the three-phase fault, where all phase voltages reverse, only memorizing of the prefault polarizing voltage can be used to achieve a correct directional determination. Normally, in distance protection, the memory voltage is used only when the voltage is reduced to some percentage of the nominal voltage. This criteria cannot be used when a voltage reversal occurs. The use of the memory voltage must be controlled by general nondirectional three-phase fault criteria.

The time the memory voltage can be used must be limited to approximately 100 ms. Today, memory can be made very accurate, but in a three-phase fault, the prefault condition should be extrapolated for only a limited time

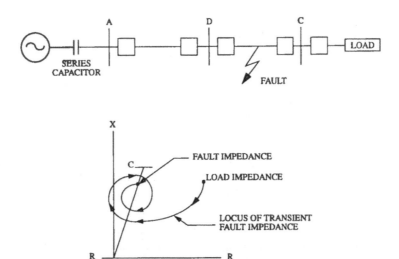

Figure 7-5 Transient fault impedance.

after the fault. The network is in a changing state and will lose synchronism with the memory. Therefore, directional measurements have to be sealed-in before the memory becomes unreliable.

The low-frequency transient that occurs in a series-compensated system can be seen in the impedance plot as a transition from load impedance along a logarithmic spiral, such as that shown in Figure 7-5. This transition can cause both over- and underreach as well as a false direction decision. This problem is overcome with high-pass filtering of the measuring quantities.

The transient caused by flashing of bridging gaps will jeopardize the security of the relaying system. Also, line-energizing transients are high frequency and could cause some relays to operate. To avoid unwanted tripping, low-pass filtering of the measuring quantities is necessary.

The foregoing problems require that band-pass filtering be used on the measuring quantities. The requirement of band-pass filtering exists in all distance protection, but is much more pronounced in applications involving series-compensated networks, to avoid unwanted operation.

With increased current through a capacitor bank and increased "conducting angle" of the parallel metal oxide element, the capacitive reactance will start to diminish and the combination will have a resistive component, as seen in Figure 7-1. When setting impedance relays on the compensated line, allowance for this apparent resistance is necessary to assure tripping at all fault current levels.

The measured impedance during a fault on the compensated line will change owing to the negative reactance of the capacitor and the status of the gaps and metal oxide element in parallel with the capacitor. To overcome this change in impedances, a zone-1 impedance measurement must be shorter than the line with unabridged capacitor, and an overreaching zone must be longer than the uncompensated line. With this setting the *permissive underreaching* transfer trip system cannot be used. Of the distance-type schemes only overreaching systems, such as permissive overreaching transfer–trip (POTT) or a blocking system may be used on series-compensated lines.

The voltage reversal on the bus for a fault on a series-compensated line can be seen as a false fault on adjacent lines. On these lines, the voltage is reversed at one of the terminals, and the voltage is "zero" in one position on the line. An unwanted trip is avoided by using the proper polarization, as previously described.

At the terminal remote from the bus where the voltage reverses, the voltage zero cannot be distinguished from a real fault. Therefore, independent zone-1 trip can generally not be used in such a location.

5. DIRECTIONAL COMPARISON PROTECTION

The direction comparison scheme consists of directional relays with limited or infinite reach. With measuring elements similar to distance relays, the directional comparison protection will have the same limitations as described earlier for distance protection. To overcome some of the limitations, negative- and zero-sequence directional relays are used for unsymmetrical faults, together with a directional relay of the impedance type for three-phase faults. Both negative- and zero-sequence relays, polarized with their respective voltages, will respond with the directional sense, shown in Figure 7-6 for ground faults, in approximately 15% of the line lengths near the capacitors in systems with parallel-compensated lines. The directional comparison protection will thus fail to operate for these unsymmetrical faults and, furthermore, this scheme has the same limitations as distance protection for three-phase faults.

The directional comparison type concept is also affected by the well-known problem of power reversal illustrated in Figure 7-7. Tripping at the right-hand terminal near the fault, produces a reversal in the upper line. The power reversal is overcome by transient blocking logic, but this may introduce a delay in clearing an evolving fault that was initially external and then became internal. Transient blocking is also required for the POTT system. The performance of the directional comparison scheme is relatively similar to the distance protection scheme, but is lacking the backup function inherent in the distance protection scheme.

6. DIRECTIONAL WAVE PROTECTION

Directional wave protection is based on the use of directional detectors that evaluate the sudden change in voltage and current caused by a fault. The directional discrimination is accomplished in 2–4 ms. During this very short measuring time, the voltage across the series capacitor bank will have changed very little. To change the capacitor voltage, energy must be forced into the bank. The inductances in the system will limit the amount of energy that can be transferred during the first few milliseconds.

Because the capacitor bank voltage is not changed significantly during the time for directional measurement, the capacitor will not influence the measurement. The directional measurement is used in a directional comparison co-operation scheme by linking the two line terminals by a single-transmission channel. Directional wave protection will thus not be affected by the series compensation, and can be used on compensated lines as well as adjacent lines

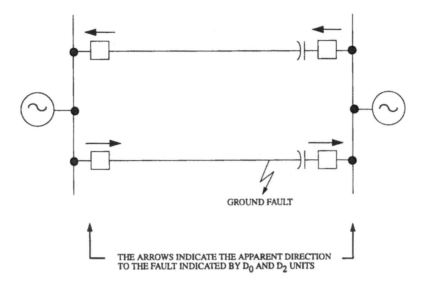

Figure 7-6 Negative-4 and zero-sequence directional relays.

that are affected by the compensation. The sensitivity is increased by weak–end-infeed echo and trip functions.

The bridging of a capacitor on a protected line will be seen by directional wave protection as a change similar to an internal fault. The sensitivity will be set not to trip when the capacitor is bridged for an external fault. This sensitivity is sufficient for all phase-to-phase faults including three-phase faults.

To increase the sensitivity to ground faults, an additional directional detector with higher sensitivity can be

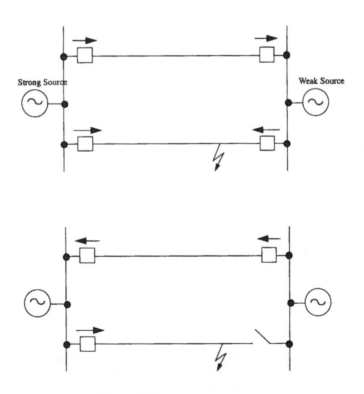

Figure 7-7 Directional reversal.

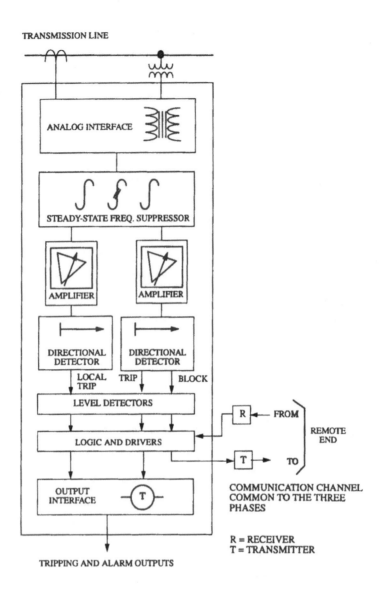

TRANSMISSION LINE

R = RECEIVER
T = TRANSMITTER

TRIPPING AND ALARM OUTPUTS

Figure 7-8 Directional wave protection block diagram.

included. To avoid false tripping at bridging of the capacitor bank, an additional criterion is required. This criterion is a control that a zero-sequence current must exist 20 or 40 ms after the operation of the directional wave detector. A system block diagram is shown in Figure 7-8.

7. PHASOR DIFFERENTIAL

The phasor differential system is similar to a phase-comparison system except that the relative phase position of currents at the two ends of a transmission line is not the sole criterion used by this relay system for judging whether a fault is internal or external to the line. Rather,

the actual phasor differential between the local and remote quantities, along with a summation restraint, is used to establish the need to trip or restrain.

One form of phasor differential, the LCB-II (or REL356) system, utilizes a single-phase voltage that is developed at each transmission line terminal. This voltage is proportional to the symmetrical component content of the input currents. A different selectable weighting-factor is used for each of the positive-, negative-, and zero-sequence components. The resultant voltage is referred to as the filter output voltage.

Communications between the two ends of the transmission line is achieved in all of the conventional ways. This fundamental frequency voltage (50 or 60 Hz) that is developed is sampled by a pulse period modulator. This device creates a voltage having a frequency that is inversely proportional to the instantaneous magnitude of this voltage. The nominal frequency about which the variations are generated is 1700 Hz. A single voice band in each direction is adequate.

With the frequency of the communications signal being dependent on the instantaneous value of the symmetrical component filter output voltage, the receiving terminal is able to convert this readily into an analog voltage, accurately recreating the remote filter voltage. Each terminal then has access to its properly delayed local filter output voltage and the remote filter voltage. A simple comparison, accomplished in a manner similar to that for a generator differential application, is then made. The operating quantity that is used is the magnitude of the phasor sum of the two filter voltages. The restraint quantity is the sum of the magnitudes of the two.

This scheme is suitable for use on some lines equipped with series capacitors. Voltage is of no concern to this system. External faults produce no problems. Internal faults with outfeed that does not cause gap flashing may be a problem, although 20% outfeed can be accommodated for an internal fault. High-resistance internal ground faults with large "through-load" current may produce a problem. A simultaneous open and internal ground fault or a simultaneous external and internal fault may be troublesome. Unsymmetrical gap flashing or incomplete transpositions may cause three-phase faults to go undetected, owing to heavy zero-sequence weighting.

For the example in Figure 7-4, the large *outfeed* on line 1 at B would cause this comparison to fail. However, the very large current at B is well over the level that could be used for a "high-set" trip. This would trip 2 directly and clamp the channel, allowing tripping to take place at 1.

8. PHASE-COMPARISON

The basic phase-comparison system is a scheme in which a single-phase filter voltage is developed at each transmission line terminal that is dependent on the symmetrical component content of the phase currents. Communications equipment is keyed during the positive half-cycle. By using the received signal, a comparison is then made to establish whether the filter voltages are approximately coincident, as they would be for an internal fault, or are 180° apart for an external fault or load condition; unlike the phasor differential scheme, *no* outfeed can be accommodated by the phase-comparison system for an internal fault.

Fundamentally this scheme has served the industry well, but it is subject to all of the problems listed for pha-

sor differential plus the problem of light current crossover for low-magnitude external faults. Neither has inherent backup nor provision for single-pole trip.

9. OFFSET KEYING

The simplicity of the phase-comparison protective relaying system encourages its consideration for series-compensated transmission line applications. With no voltage used for the basic relaying concept, all of the problems associated with voltage aberrations are eliminated. Any "through" phenomena, being at rated frequency, low-frequency, or high-frequency, create equal influences at each of the line terminals. Only the charging current of the line exists as a quantity with which we must reckon to assure security. This is easily handled by the offset-keying concept as will be seen later.

For internal faults, two circumstances that must be handled are (a) outfeed at one terminal and (b) large through-load current, with a high-resistance internal ground fault.

Combining segregated-phase comparison with offset-keying produces an excellent combination for single-phase trip, series-compensated line relaying. Segregated-phase comparison systems compare the "in" and "out" currents for each phase at the two terminals of the transmission line, and do not use a three-phase composite symmetrical component filter. The system becomes inherently phase selective and may be used to trip only the phases involved in a fault, irrespective of the number of phases, if desired. It is oblivious to the usual difficulties encountered with faults involving more than one transmission line. Evolving faults produce no problem. Zero-sequence mutual has no detrimental effect.

A sophisticated communications system is required if all of the pertinent information needed from the remote terminal for a trip decision at the local terminal is to be transmitted over a single voice channel. The REL 350 system accomplishes this using a quadrature–amplitude-modulated system. One voice channel total (for three-phase and one-ground subsystem) is required in each direction, and the coded signal includes the information required for another function, such as remote trip.

Figures 7-9 and 7-10 describe the offset-keying concept applied in the REL 350 segregated–phase-comparison system. An important characteristic of this system is that three distinct current levels are established; the channel (trip-positive) keying level, the local positive, and the local negative levels. The keying level is positive and set comfortably between the other two in magnitude, with a margin to accommodate charging current. The channel is

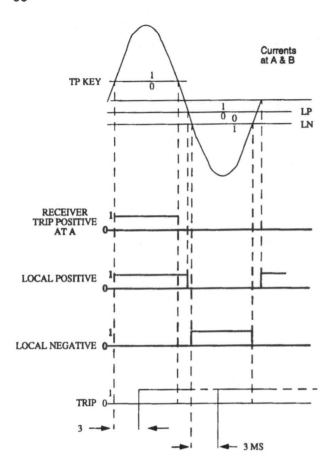

Figure 7-9 REL 350 waveforms internal fault, both currents.

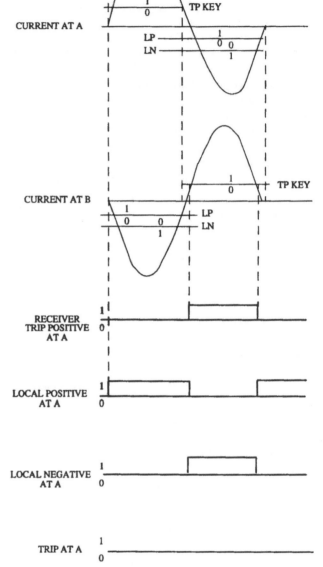

Figure 7-10 REL 350 waveforms external fault.

keyed to the trip-positive state when instantaneous current is above a certain positive threshold, typically 3.579 A (3-A RMS for 3-ms key). Local positive is at a "1" state for instantaneous current more positive than a negative-threshold value, typically 2.386 A. Similarly, local negative is at a "1" state for instantaneous current more negative than a negative-threshold value, typically 4.772.

For those power system conditions that produce outfeed (one terminal feeding current into the fault and the other terminal current 180° out-of-phase with that) for a fault internal to the transmission line, the TP key level is set above the magnitude at the outfeed current location. Tripping occurs as a result of a large local negative value at the other terminal. All tripping (including a high-set overcurrent trip) clamps the transmitters to the trip-positive state.

Previous implementations of this concept have utilized four frequency-shift voice channels in each direction and contained no distance backup. The REL 350 system is a coded quadrature–amplitude-modulated system and requires only one voice band in each direction for the four subsystems (three-phase and one ground). Further, a two-

zone, time-delayed, offset-distance, nonpilot backup system may be incorporated as an inherent complementary function. Phase selection utilizes a patented $I_1 + I_2$ comparison scheme for optional backup single-pole tripping. Dual blinder out-of-step tripping and blocking is also included.

All of the shortcomings of the fundamental frequency–voltage-dependent systems have been overcome by the

REL 350 system, and exhaustive tests indicate that no new ones have been created.

10.   IMPORTANCE OF COMBINATIONS

Historically, it has been demonstrated that any problem we can identify and repeat, we can find a solution for. Problems have a way of lingering, unidentified until the worst possible moment arrives. Sooner or later every protective relaying system displays a shortcoming. Some fault combination, some fault incidence angle, some set of system parameters, some prefault load condition, some phase-impedance unbalance, some unsymmetrical gap-flashing event, some switching condition, some off-normal frequency excursion, or some instrument transformer malfunction will cause a deficiency to appear. It may be a large one or a small one.

By using a second protective-relaying system having a different measuring concept (even if it only involves a phase roll) will invariably cover that rare exposed shortcoming of the first system. A second different system is recommended in any important transmission line application, particularly where a series capacitors or single-phase tripping is used. These produce very complex power systems, and they generate a host of challenging phenomena.

11.   IMPORTANCE OF MODEL POWER SYSTEM AND EMTP TESTING

A modern model power system (MPS) is extremely expensive if it is to fulfill the needs for elaborate and accurate power system representation. What this large investment allows is very wide flexibility in parameters, refined control of fault application, reproducible power system phenomena, wide variety of load conditions, extensive data acquisition and high-resolution, and clear plotting capability.

The electromagnetic transients program (EMTP) adds to this important test facility, the ability to model a far more extensive system, to do so with uniformly distributed parameters if desired, to totally define the waveforms, and to eliminate undesired component effects. Provision for both types of tests must be incorporated in a manufacturer's developmental facility. Each has its strengths and weaknesses.

The MPS is most likely to provide useful surprises in the behavior of a protective-relaying system. Being a real-time system, with real-current transformers that have an arbitrary residual flux level, with real trip and reclose action, arbitrarily generated control power transients and realistic inequitable pole action of breakers, testing; occasionally leads to unexpected performance of the relays. However, all of this can be repeated on the MPS.

From the viewpoint of ease of generation of the next case, infinite ability to explore the fringe phenomena, and feasibility to examine intricacies on virtually a microscopic basis, the MPS has no equal. The immediate interaction between the MPS, the relaying system, and the test engineer on a closed-loop basis provides a synergism that is not possible in any other way.

The EMTP approach, on the other hand, allows a pre-established sequence of waveforms unaffected by current transformer (ct) resistance, unaltered by undesired ct and ccvt transient behavior. It is effective for users and small manufacturers, with limited testing facilities, who desire the assurance provided by the application of a fixed series of difficult tests.

A comparison of the programming requirements is useful. Other than the preliminary one-time programming for the control and data collection, none is required for the MPS. New cases can be set up in minutes by personnel with no programming knowledge. Each new case to be run on EMTP requires the attention of an experienced programmer.

Both facilities provide important capabilities not present in the other, and it is expected that both will continue to be used in the development of protective relays.

12.   CONCLUSIONS

This chapter has attempted to describe the difficulties associated with the relaying of a transmission line equipped with a series capacitor and has described various transmission-line relaying systems.

The segregated-phase comparison system and the directional wave protection systems appear to offer the most straightforward solution to this difficult relaying application.

8

Single-Pole Tripping

WALTER A. ELMORE

1. INTRODUCTION

Historically, the relaying equipment associated with single-pole tripping has been complicated, expensive, and space-consumptive. Although the root technology is still complicated, the incremental expense over three-pole tripping is small, and the panel space requirement is the same as that for three-pole tripping in modern single-pole relaying systems.

Another deterrent to single-pole tripping has been that breakers have traditionally been equipped with three-pole mechanisms. Today EHV breakers, owing to the very wide pole separation, have been forced to use separate mechanisms, even for three-pole applications, and independent pole-operated breakers are now available from 72.5 to 800 kV (though with considerably higher cost when less than 345 kV).

This chapter describes (a) the advantages of single-pole tripping, including stability considerations; (b) phase selection methods to assure proper faulted-phase identification; (c) the difficulty of subsequent fault identification, as for example, a "B–G" fault after tripping for an "A–G" fault; (d) the symmetrical component representation of a system following single-pole tripping; (e) the influence on rotating machinery of the transmission line single-phasing, following single-pole tripping; and (f) the near-certainty of increased use of single-pole tripping as a result of decreasing availability of rights-of-way.

2. SINGLE-POLE TRIPPING CONCEPT

Single-pole tripping, often called, somewhat erroneously, single-pole reclosing, provides some interesting benefits, but also some technical challenges. The strategy for single-pole tripping is to isolate only the faulted phase at the occurrence of a transmission line single-line-to-ground fault, and to isolate all three phases for all other faults. The benefit to power system stability is obvious when we consider that the two system segments, which the transmission line interconnects, remain metallically interconnected by the two unfaulted phases during the single-phasing period, and as a result, a substantial amount of synchronizing power can flow. Also, the influence of voltage variation throughout the power system is reduced as a consequence of single-pole tripping.

Following a dead period of sufficient time for the arc to become deionized, the open pole is reclosed. For temporary faults, the circuit is restored with relatively minor effect on the power system. For more persistent line-to-ground faults, the circuit breaker is then tripped three-pole, with subsequent action being similar to conventional tripping and reclosing, as dictated by the particular utility's

99

practice, or to allow the two systems to remain separated until resynchronized.

3. DIFFICULT PROBLEMS

Although intuition may indicate that the selection of the faulted phase should be a simple matter, it is not. Fault current in one phase only, is obscured by the presence of load current. Load current may add to or subtract from the fault current and, certainly, a balanced load will produce current in the unfaulted phases. Also inequitable distribution of symmetrical-component sequence quantities will generate unfaulted-phase currents for a phase-to-ground fault even though there is no load current. Once a fault is identified as being on a particular phase, and single-pole tripping has been initiated, there will be a strong tendency for relays to trip incorrectly as a result of the unbalanced-loading condition that remains. Zero-sequence and negative-sequence currents will flow in the protected circuit and, perversely, in such a manner that undesired operation of directional units will be caused using those sequence quantities *at both transmission line terminals.*

At the same time, the relaying system must be able, during the single-phasing period, to recognize a fault that occurs on one of the sound phases and to initiate three-pole tripping at high speed. Also during this period, it must be secure against false tripping in response to all varieties of external faults or severe swing conditions.

To accomplish all of this requires a high order of relaying refinement and serious attention to all of these constraints.

4. EQUIVALENT DIAGRAM

The symmetrical component representation of a system, with a line operating with one pole open after single-pole tripping, is shown in Figure 8-1. X_{1S}, X_{2S}, and X_{0S} are equivalent source positive-, negative-, and zero-sequence impedances; X_{1L}, X_{2L}, and X_{0L} are line impedances; and X_{1U}, X_{2U}, and X_{0U} are component impedances beyond the open. Figure 8-2 shows a reduction of Figure 8-1 to two machines and a single interconnecting reactance per phase. The power transfer associated with the open-phase condition is determined by the equation shown in Figure 8-2.

5. STABILITY

Single phase-to-ground faults do not seriously impair the ability of a transmission line to carry power, but 20% re-

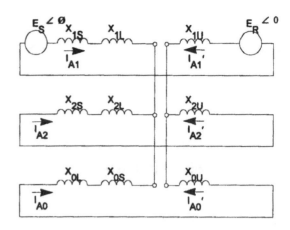

Figure 8-1 Symmetrical component network interconnection for open-phase condition.

duction in the power transfer curve is typical. The tripping of three poles in a complex network may slightly hamper the power-transfer capability (particularly if the source/line impedance ratio is large). However, in applications involving the interconnection of a generating plant to a power system through a single transmission line, single-pole tripping may be the only way to maintain synchronism of the plant with the system, and then only for temporary single line-to-ground faults.

Figure 8-3 shows the substantial influence on stability that single-pole tripping can provide when one interconnecting line exists between a generating plant and a power system, or between two power system segments. Figure 8-3b describes the trajectory on the power swing curves from normal to that with a phase-to-ground fault. When three-pole tripping occurs, all synchronizing power is lost until reclosing takes place. If reclosing is successful, swinging continues. If areas 2 + 4 above the original power line can equal areas 1 + 3 below the line, stability may be possible. Much depends on trip time, the H constants, the actual levels of accelerating and decelerating power, reclosing time, and the actual system reactances.

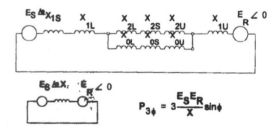

Figure 8-2 Reduction of Figure 8-1.

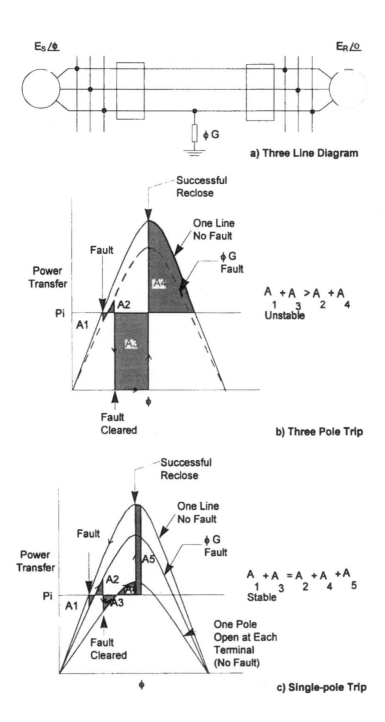

Figure 8-3 One interconnecting circuit.

Figure 8-3c describes the similar process accompanying single-pole tripping. The likelihood of retaining stability with this strategy is very much greater. The obvious benefit that can be seen in this diagram is from the vast reduction in the accelerating power to which the sending end is subjected during the single-phasing period.

These curves are only typical and are based on the assumption of convenient system reactances; comparable results occur with other reasonable parameters.

Figure 8-4 shows the circumstances accompanying a two-line case, with a fault on one. Again the power transfer curve dips to approximately 80% of the nonfault level.

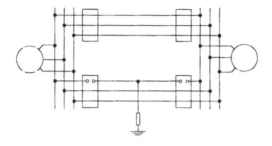

Three Line Diagram

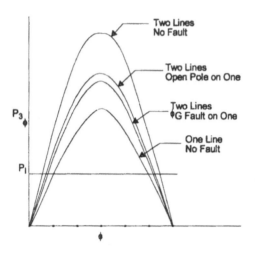

Power Transfer Curves

Figure 8-4 Two interconnecting circuits.

With single-pole tripping of the faulted circuit, the power-transfer curve is elevated only slightly when that pole clears. Whether single-pole tripping is used or not, stability is never in question when only single line-to-ground faults are considered; however, it must be recognized that with one line out of service for maintenance the system and the stability considerations revert to the single-line case of Figure 8-3.

6. SELECTIVE POLE RELAYING

Although single-pole tripping is considered as particularly beneficial for systems, such as that of Figure 8-3 or of those of Figure 8-4, which can degenerate to that of Figure 8-3 during an outage, some thought might be given to se-

lective pole tripping. Selective pole tripping would isolate only those phases that are involved in a fault. This would extend the benefit previously described for single-pole tripping to phase-to-phase faults (and phase-to-phase-to-ground faults if desired). For a $\phi\phi$ fault, for example, only the breaker poles associated with the fault are tripped. Actually, only one pole at each end of the transmission line need be tripped to clear a temporary fault. Phase selection would be of no difficulty whatever using segregated phase relays (such as REL 350). Tripping two poles at each end for a $\phi\phi$ fault would require only minor adjustment of the standard scheme. Roughly 20% of the original power transfer capability is retained for two-phase tripping, compared with zero when all three poles are tripped in Figure 8-3. With all combinations of pole tripping, close examination of circuit breaker recovery voltage characteristics would have to be made. Standard breakers are compatible with single-pole tripping, but they may not be for some types of selective pole tripping.

7. PHASE SELECTION

The proper selection of the faulted phase, or phases, on a protected transmission line is imperative if the relays are to perform their function correctly. Proper operating and restraining quantities must be paired to properly identify fault type.

Many methods have been used for identifying fault type, some more successful than others. One obvious implementation uses individual phase overcurrent. Overcurrent in one phase would presumably indicate a phase-to-ground fault: in general, it does not. The use of overcurrent for faulted-phase identification has severe shortcomings. Unequal symmetrical component distribution or a large load component can produce substantial current in the unfaulted phases, obscuring the character of the fault. The use of distance relays has some appeal, but may fall short because of the tendency for more than one phase unit to operate for some single-phase-to-ground faults.

One scheme that has been used successfully for identifying the faulted phase involved in a line-to-ground case compares the phase relationship between an individual phase-negative sequence current and that of zero-sequence current. This scheme, without help, falls short (Fig. 8-5). For example, a phase-A negative-sequence current aligns with zero-sequence current for a phase-A-to-ground fault. Unfortunately similar alignment occurs for a B–C ground fault. This can be overcome by the patented scheme shown in Figure 8-6, which adds to the logic a phase-shifted voltage (shifted by a small angle to accom-

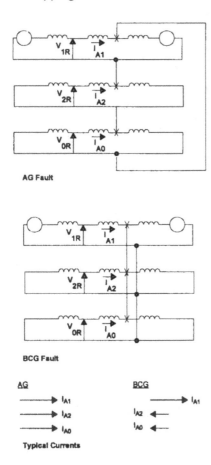

AG Fault

BCG Fault

Figure 8-5 I_{A2} and I_{A0} for AG and BCG faults.

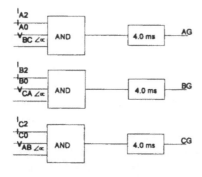

Figure 8-6 One form of phase selector.

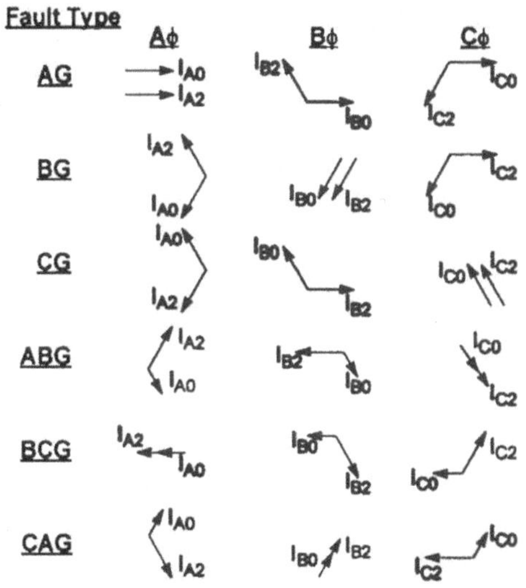

Figure 8-7 Relationship of negative- and zero-sequence components for ground faults.

modate the angle of lag expected for ϕG faults). The 4.0 ms coincidence timer imposes the requirement that all three phasor quantities in the comparison fall within a 90° band. This minimizes the likelihood of a false identification of fault type with a high value of ground fault resistance. Figure 8-7 shows the relationship between I_0 and I_2 for various fault types.

Another variation of this scheme for faulted-phase selection also places dependence on I_0 and I_2 coincidence for each phase, but excludes $\phi\phi$G faults by identifying them in terms of the extent of voltage reduction in the individual phases and blocks tripping. The blocking of ground elements places partial dependence for $\phi\phi$G tripping on the *phase* relays, requiring in some instances, special compensation (such as $I_A - 3I_0$) for the phase elements designated to detect 3ϕ faults.

Another important scheme utilizes $I_{A1} + I_{A2}$. These two quantities are well behaved relative to one another and, for a phase-to-ground fault, resistance influences neither their phase nor magnitude relationship. Prefault load flow can, however, affect the positive-sequence current phase and magnitude. To eliminate this effect and to simplify the process for microprocessor implementation, positive-sequence current, produced by the fault, plus negative-sequence current can be obtained by:

$$I_{A1} + I_{A2} = I_A - I_{AL} - I_{A0}$$
$$I_{B1} + I_{B2} = I_B - I_{BL} - I_{B0}$$
$$I_{C1} + I_{C2} = I_C - I_{CL} - I_{C0}$$

This comes about because I_A, for example, is total current during the fault including the prefault load current, and also the fault component of I_A is $I_{A1} + I_{A2} + I_{A0}$. Removing the prefault load current and I_{A0} from the measured phase current during the fault leaves only $I_{A1} + I_{A2}$, the

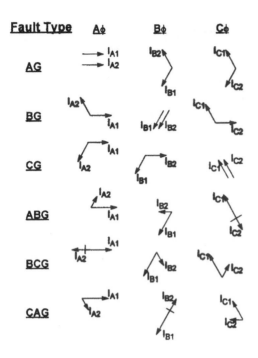

Figure 8-8 Relationship of positive- and negative-sequence components for ground faults.

positive- and negative-sequence fault components of A phase current. Comparing the magnitudes of the RMS values of the three-phase component sums, the type of fault and phase(s) involved is apparent (Fig. 8-8). The criterion used is based on a comparison of the phasor sum with 1.5 times the phasor sum of the other two phases. For example:

$$|I_{A1} + I_{A2}| > 1.5\ |I_{B1} + I_{B2}|$$

and

$$|I_{A1} + I_{A2}| > 1.5\ |I_{C1} + I_{C2}|$$

identifies a "phase–A"-to-ground fault.

$$|I_{B1} + I_{B2}| > 1.5\ |I_{A1} + I_{A2}|$$

and

$$|I_{C1} + I_{C2}| > 1.5\ |I_{A1} + I_{A2}|$$

identifies a BC or BCG fault.

This method has the interesting quality of utilizing symmetrical component quantities without having to generate any but zero-sequence current and having that available from the sum of the individual phase currents.

From Figure 8-8, it is apparent that comparison of these sums will allow a clear identification of a phase-AG fault, for example. Similar comparisons of other fault types

show that all of them are identifiable using this method except the three-phase faults, which contain no negative-sequence current (except in the practical case involving unequal phase impedances, for which the level would be expected to be quite small).

The absolute method for identifying the type of fault and which phases are involved utilizes segregated phase comparison. The currents at each end of the transmission line are compared by a communication channel *for each phase*. Current-in equals current-out for a sound line. Any difference other than that caused by uncompensated; distributed capacitance is fault current. Even "cross-country" faults, those involving more than one three-phase transmission circuit, are clearly identified as to which phases are involved and on which circuits they exist in spite of the presence of distorted voltages for one circuit with a fault on the other. Voltage is not used in this scheme. The need for single-pole or three-pole tripping is clearly identified immediately without the complexity or time delay of other methods. Essentially simultaneous tripping occurs at both transmission-line terminals even though a strong source exists behind only one of them.

8. DURING SINGLE PHASING

Following single-pole tripping, a small secondary arc will be sustained as a result of capacitive current flowing between the still-energized sound phases and the now disconnected faulted phase. This current causes the deionization time to be prolonged beyond that which would be needed for three-pole tripping, possibly as much as five to eight cycles. Reclosing time must be extended by an appropriate amount to accommodate this.

During the short single-phasing period, positive-, negative-, and zero-sequence currents flow in the line (in one end and out the other) and the inherent scheme or backup devices must be chosen carefully to avoid misoperation, while at the same time being able to recognize subsequent faults that may then occur involving the "sound" phases. This becomes an awesome task for most relaying schemes, because of the unbalanced load influences, possible line-side location of voltage transformer, all possible combinations of internal and external faults, and the necessary blocked character of some relaying functions. The remarkably simple character of segregated-phase-comparison relaying is confounded by *none* of this.

Adjacent circuits also experience the short-term influence of single-phasing, and the relaying associated with those circuits must either ignore this effect, through the nature of the relaying or its setting, or it must be undesirably time-delayed long enough for successful reclosing or three-pole tripping on the faulted circuit to take place.

For long transmission lines (for example, more than 45 miles of 500 kV; [1]) capacitive and inductive coupling to the de-energized phase from the two energized phases may be severe enough to prevent deionization of the fault arc. To mitigate this, four-legged high-voltage reactors may be required to provide a near-parallel resonant supply to the arc, producing a self-quenching effect [2].

The switching action associated with single-pole tripping produces swinging between machines, as the variation of φ in Figure 8-3c shows. Blocking of distance units during these swings may be necessary to avoid undesired response. Where line-side voltage transformers are used, the relaying system must take cognizance of the need to block tripping during swings, even though one voltage transformer may be deenergized.

Rotating machinery experiences exaggerated heating effects as a result of the presence of negative-sequence current flow. The closer the proximity to the open point, the greater will be this influence. The relaying system must be equipped with provision for limiting the single-phasing period to minimize the exposure of rotating machinery to this hazard.

9. TREND

The continuing restrictions on right-of-way availability and the resultant tendency toward increased transmission line loading forces reevaluation of practices that may provide even marginal improvements. In some cases very large benefits in transient stability are possible using the single-pole-tripping concept. In other cases, only marginal improvement is possible. The key thing that makes single-pole tripping a viable factor is the minor increase in relaying space and cost and the minor, if any, increase in circuit breaker costs that are involved in such an application, contrasted with what was involved only 20 years ago. Also, the relaying available is highly sophisticated and incorporates the accumulated years of experience, while requiring only additional built-in output relays to provide single-pole tripping.

10. CONCLUSION

The relaying and circuit breakers are available for single-pole tripping, allowing, with minor cost addition, the possibility, in some cases, of substantial improvements in power system stability. The problems are well defined and the solutions, through many years of experience, are clear. The cleanest and least-complicated relaying system appears to be some form of segregated-phase–comparison.

REFERENCES

1. EW Kimbark. Suppression of ground-fault arcs on single-pole switched EHV lines by shunt reactors. IEEE Trans Power Apparatus Syst March 1964.
2. IEEE Committee Report. Single phase tripping and auto reclosing of transmission lines. IEEE Trans Power Deliv 7(1), 1992.

9

Substation Automation and Relay Communications

WILLIAM J. ACKERMAN

1. INTRODUCTION

There is still no industry-approved definition of substation automation. This chapter will attempt to provide a general review of what constitutes substation automation and, then, define some of the forces that are driving the requirements for automation. A key tool in the automation of a substation is the use of intelligent electronic devices (IEDs). The use of IEDs presents several implementation considerations, including those of protocol and compatibility with existing substation equipment. There appears to be a distinct difference between systems that provide interfacing functions, and those that are capable of providing many true automation functions. Some of the differences in these types of systems are identified, as are potential candidates for automation. Modern substation automation systems also have a significant influence as a corporate data source outside the substation. There are significant unanticipated consequences that can result from the use of IEDs, some of which are identified and discussed.

A key element in any substation automation effort is that of communications. Two areas must be considered: communications between and within the substation, and between the substation and the "outside world." Here we will take outside world to represent other substations, corporate offices, other control and monitoring systems, and in some cases, the general public.

Implementation of substation automation can have a substantial effect on other utility data-gathering and control systems and data storage and analysis systems. For convenience, they are referred to in this chapter as supervisory control and data acquisition (SCADA) systems, energy management systems (EMS; generally applied to facilities used to monitor and control the transmission system), distribution management systems (DMS; generally applied to facilities used to monitor and control the distribution system), or simply information management system (IMS; which we will apply to any system that gathers, uses, stores or processes data, and perhaps enables people or computer systems to exercise control over some aspect of the power system).

2. SUBSTATION AUTOMATION AND DRIVING FORCES

Just about every substation is already automated in some respect, or by someone's definition of automation. The capability for automatic reclosing after a breaker operation is a common automation function. In fact, many transmission or bulk power substations are perceived as being automated because of the use of a remote terminal unit (RTU) to monitor the substation real-time parameters through SCADA. Some typical automation functions in-

107

clude automatic tap-changing, automatic failover to alternative lines or sources, transfer tripping, corridor monitoring and control, or others. On the other hand, most distribution substations are in the early stages of automation and, in fact, many are not equipped with RTUs or with the capability for SCADA. The use of microprocessors, programmable logic controllers (PLC), embedded systems, and other developments of modern technology opens up a whole new meaning to substation automation. The automatic functions are typically more complex, include a greater variety of contingency conditions, and respond faster than the simple-relaying schemes that were once considered to be "automation."

Some of the driving forces for increased substation automation are the following:

1. Reduced costs
    -   Utilize equipment as close as possible to ratings, defer replacement costs
    -   Reduce periodic maintenance visits to the substation
    -   Switch to reliability centered maintenance (RCM); that is, perform maintenance as it may be needed, not simply because it has been scheduled at some periodic interval
2. Improved efficiency
    -   Closer monitoring and control of voltage to customers
    -   Reduce or eliminate equipment overloads on the power system
    -   Maintain a high level of power quality
    -   Maximize asset utilization without impairing overall power system security and reliability
3. Improved reliability
    -   Automatic switching and restoration sequences to reduce outage time
    -   Dispatcher switching for faster recovery from outages
    -   Reduced vulnerability to equipment outages

Power system operators require the most modern analytical tools, running on powerful computers, to accomplish these ends. Many of the programs are already available or are in advanced stages of development. However, the usefulness and accuracy of every single program is totally dependent on the quality and timeliness of the data coming out of the substation. Clearly, if the average substation-metering data has an accuracy of ± 5%, then the results of any analytical program using that data cannot be much better. A general increase in data accuracy by one or two percentage points, which is now easily achievable with IEDs, can result in an increase (or decrease) in transmission capacity on the same order of magnitude. De-

ferring (or requiring) a transmission capacity increase of just 1 percentage point has substantial capital expenditure consequences.

In the business world, there is a need to provide clear and unambiguous decisions. That is: Yes—a proposed transaction is acceptable; or no—it is not. However, in the world of engineering and power system operations, the decisions are seldom pure black or white. In particular, equipment-operating limits are generally "soft," rather than hard. Thus, we see apparatus having multiple ratings and alarm limits based on long-term, short-term, and maybe, even emergency considerations. These multiple limits are because, for example, exceeding a transformer nominal rating does not immediately destroy the transformer, it just reduces the transformer's expected life. The same is basically true of most other power system equipment. Depending on weather and other conditions, exceeding the nominal rating may have no effect whatever on the apparatus. One response to this problem will be the installation of condition-monitoring equipment on most power system equipment in the substation. This may well be in the form of another set of IEDs, particularly if there is a requirement for maximum possible accuracy. Generally, the condition-monitoring IED will supply a significant quantity of "raw" data that will require sophisticated software to perform the analysis function so that preventive maintenance can be done on the basis of need, rather than on an arbitrary schedule.

It must also be noted that if replacement costs, lifetime reduction costs, and "wear-and-tear" costs are to be factored into service charges, then the charging utility will be required to have a comprehensive database that can be used to develop and justify such charges. Thus, automating the substation to obtain the required data will become a necessity.

3. BENEFITS OF INTELLIGENT ELECTRONIC DEVICES AND IMPLEMENTATION CONSIDERATIONS

The key to substation automation has to be communications technology. Early SCADA systems operated at 1200 bits/second (bps) or less, and required conditioned voice-grade lines for data transmission. Costs of leased lines included both a fixed monthly charge and a monthly per-mile charge; and could approach $1000/month. As a result, only transmission substations were equipped with RTUs because bulk power system reliability demanded information, regardless of cost. Great emphasis was placed on development of protocols with high efficiency (high percentage of "useful" information bits transmitted vs. total bits transmitted) and reasonable security.

As communications capabilities increase in the form of modern modems, operating at higher bit rates, and in the form of many more options for communications media (copper lines, radio, fiber optic, satellite, and such), it becomes possible and economical to push SCADA and automation deeper and deeper into the power system. In fact, many utilities are now considering, and some are implementing, direct communications circuits right to the individual residential customer's home.

The benefits of modern substation automation are applicable to both new and existing sites. The most significant benefits for existing sites are the improved quality of information and the elimination of obsolete equipment. Overall, there is also a benefit in standardization of equipment and material. Standardized configurations help control training costs and ensure that all personnel can safely perform their work in and around the substation.

The major benefit for new construction is the lower installed cost that results from savings in construction (no cable trenches, covers, trench drains, or the like) and elimination of massive amounts of physical wires and cables. In many installations the cables and wires are vulnerable to damage from water, rot, and rodents, presenting significant maintenance and replacement costs. Because less cable is used, the added expense of making what cable is used more resistant to damage is easily absorbed.

The use of IEDs in a substation also presents significant savings. Consider, for example, the current practice as relates to the installation of RTUs. Typically, the utility will use a transducer, or multiple transducers, connected to the PTs and CTs to obtain a $0- \pm 1$-ma signal proportional to the measured value (watts, vars, volts, amperes). This signal is then wired to the analog input card of the RTU where it is digitized and then transmitted to the SCADA master station. In addition to the one-time wiring and installation cost of each transducer, there is the recurring cost of periodic transducer calibration, linearity adjustment, and scaling; Analog input card input range-scaling, and possibly, the scaling values used in the SCADA database. All of these factors have a potential for error, thereby causing possibly significant errors in the final value placed in the SCADA database.

By taking the digitized values direct from the IED, all these potential errors are eliminated. In addition, the careful checking to ensure that the protection functions of the IED are performed correctly also ensures, at the same time and at no extra cost, that the data functions are also correctly performed.

When connecting control outputs from a RTU to the end device, it is common practice to use interposing relays to achieve the required contact ratings. These relays are costly, require additional wiring, and represent another area for reliability problems. By using the output connection of the IED for control execution, these problems are eliminated.

Finally, there are significant physical space savings when IEDs are used in a substation. Consider a very small substation consisting of a single incoming supply, one transformer, and two feeders.

With electromechanical relays and the usual complement of control switches, lights, and meters, the control house for this small substation will require space for at least four steel panels that mount the equipment (Fig. 9-1).

The control panels for the small substation, based on the use of electromechanical relays and meters, are illustrated in Figure 9-2. In 1998, the typical average, installed cost for each panel was about $30,000 (including all the equipment on the panel); therefore, the control equipment alone for the substation was about $120,000.

Now consider the same substation using modern IEDs for protection and control. The physical space is reduced from four panels to one and, interestingly enough, the 1998 installed cost of the single panel is still on the order of $30,000 (Fig. 9-3).

An added benefit of using IEDs is that additional functionality is often obtained. In this case, the following functions are included in the IEDs, without any additional physical space requirements, which were not on the electromechanical relay panels:

- Two steps of underfrequency shedding and restoration
- Two steps of overfrequency shedding and restoration
- Overvoltage protection
- Undervoltage protection
- Complete real-time metering and control capabilities, including all targets
- Historical records of faults, operations, min–max values, load profiles, and oscillography

4. PROTOCOLS AND DATA

There are significant compatibility issues associated with substation automation. The major issues between the substation and the SCADA system are protocols and data representations.

Typically, the substation automation system must include a protocol emulation that is compatible with that of a SCADA system. In other words, the protocol capabilities of the SCADA system are considered "fixed" and the substation automation system must be capable of interfacing with those capabilities. The problem can be compounded if there are multiple SCADA systems that require data. This situation arises when a utility has EMS, DMS, or other IMS. Often the various systems are of different vin-

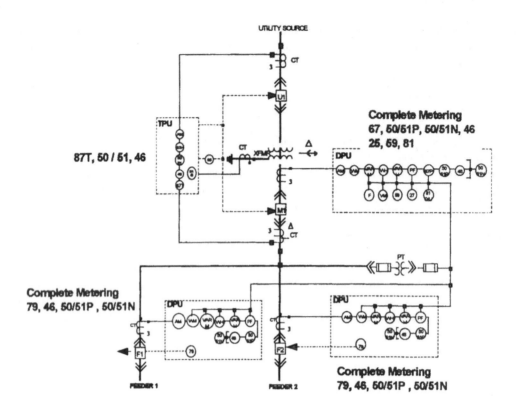

Figure 9-1 Small substation with one transformer and two feeders.

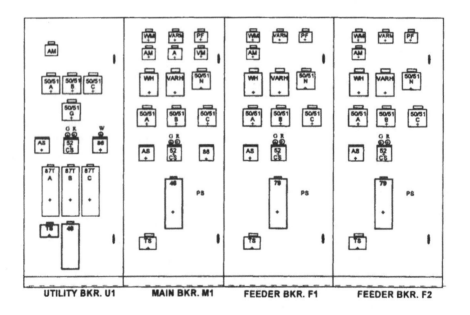

Figure 9-2 Control panels for a small substation. Typical main with two feeders using electromechanical relays—relay panel configuration front elevation.

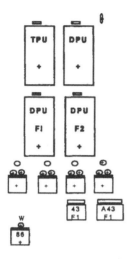

Figure 9-3 Small substation control panel using microprocessor IEDs.

tages or manufacturers, and they are probably equipped for differing protocols. Most controllers used in substation automation applications have some form of "universal protocol translator," which can be programmed to accommodate many commonly used protocols.

Database compatibility can be a problem when it comes to number representation. Seemingly minor issues can easily develop into major problems. For example, an IED communicating on a serial communications line may transmit data starting with the lowest-order bit and ending with the highest-order bit. A sign bit may precede or follow the data bits. However, the SCADA system communications interface may be programmed to expect bits in the order of sign bit followed by most-significant bit to least-significant bit order. Another typical problem is that an IED may digitize data to a precision of 32 bits. However, the SCADA system database is designed to use a maximum of 16 bits, including the sign bit. Simple truncation is usually not an answer. Or, the SCADA system is designed to accept a binary-coded decimal (BCD) number. Or the SCADA system is designed to accept a floating point number instead of an integer number. Without an intervening controller or data translator that is capable of converting between different representations, software changes in either the IED or the SCADA system will be required. Usually, either option is too costly to consider.

The Institute of Electrical and Electronics Engineers (IEEE), the International Electrotechnical Commission (IEC), the International Standards Organization (ISO), and various other bodies interested in the standards process are actively working to develop or define standard protocols-for communicating between IEDs and other equipment within a substation, between substations, and between substations and higher-level systems.

Interestingly enough, in 1998 the only officially recommended protocols (by the Power Engineering Society Substations Committee of IEEE) are DNP 3.0 and IEC 870-5. The recommendation is in the form of a "trial use recommended practice." These two protocols are closely related, and the recommendation is for their use between IEDs and RTUs. However, it should also be noted that, somewhat by default, the "standard" protocol for data transmission between SCADA masters and substation equipment is becoming DNP 3.0. This should come as no surprise, for DNP 3.0 was originally developed for use in SCADA master-to-remote communications.

The Electric Power Research Institute (EPRI) is encouraging the use of utility communications architecture (UCA) as the basis for any recommended or preferred protocol. There are several protocols that have been determined to be compatible with UCA. However, the electric power industry needs to encourage only one standard if the goal of interoperability is to be achieved. (*Interoperability* is sometimes defined as "plug-and-play," but the intent is to achieve a protocol that can be implemented by all suppliers to the industry such that all equipment can operate in harmony and without conflict). A major effort to fully define this protocol and its objects has been ongoing for some time under EPRI RP-3599 and is now continuing under the IEEE Standards Project P1625, which is assigned to the PES Substations Committee. A related development and demonstration project in the United States is the "LAN initiative." A group of utilities and vendors are working to develop a communications standard based on fiber-optic Ethernet connections. The objective of this initiative is to define a "standard" substation local area network (LAN) and associated protocol (which is intended to be UCA compatible) that will transmit data between IEDs in a substation and between substations at speeds high enough (less than ¼ cycle for up to 70 protection "commands") such that the data can be used for protective functions. At this writing, the object definitions that have been developed as part of RP-3599 are likely to become a significant part of any standard. Manufacturing messaging specification (MMS) appears to be the leading candidate for the application layer of the Open Systems Interconnection (OSI) seven-layer model, which was approved by the International Standards Organization (ISO).

Unfortunately, full definition and use of these protocols are still in the future. For installations at the present time, there are various proprietary, semiproprietary, and open protocols and interconnection methods that are used. These include INCOM, PROFIBUS, FIELDBUS, LONWORKS, MODBUS-PLUS, and others. About all that can be said about the current situation relative to standards is that there are so many different standards to choose from. It is clear that for some time to come, equipment suppliers will be required to work with many different protocols, and the using organizations (utilities) will have to pay the additional costs associated with industry support of multiple protocols.

One problem that tends to disappear when using IEDs is that of electromagnetic interference in installations. Because cabling is reduced to a few twisted pairs, shielding and isolation is more easily designed and accomplished. In any event, the use of fiber-optic communications can eliminate interference and ground potential rise (GPR) issues. Because of the clear advantages of using fiber-optic communications in the substation, more and more utilities are specifying this medium which, in turn, results in falling prices for fiber-optic facilities.

In addition to IEDs, the use of programmable logic controllers (PLCs) in the substation is becoming much more common. There are still many utilities that implement automation functions using relays, timers, switches, and such. However, the PLC allows much more complex automatic functions to be implemented. Depending on the complexity of the function being automated, programming a PLC can be a real challenge. Once the PLC is programmed, new testing techniques need to be developed to ensure the program is correctly implemented. It is relatively easy to test a PLC by applying the specified inputs and determining if the resulting output complies with the design intent. It is much more difficult, and time-consuming, to test a PLC with "unanticipated" inputs to determine if the resulting output is acceptable.

Most IEDs include a user interface that provides the same, if not more, functionality than was provided with the lights, switches, meters, and other apparatus that was mounted on the panels. However, eliminating this equipment can be "traumatic" to utility operating personnel having to learn whole new ways of operating a substation. Occasionally, this can even require changes to the utility's Safety Manual and Switching and Tagging procedures.

It must be noted that a single IED can have many more data points than a RTU for the entire substation. Thus, a single substation using IEDs for protection and control of three incoming supply lines, two transformers, a capacitor bank and 10–15 feeders might have more data points (~25,000) than a large SCADA system). As a result, databases for substation controllers tend to be standardized, such that a database can be designed once for each IED and then replicated for additional IEDs of the same type. Likewise, displays and data access techniques tend to be standardized.

Many substation automation systems still incorporate ac-operated equipment (such as a CRT), rather than incur the additional expense associated with dc-operated equipment. Thus, most systems are equipped with an inverter to operate off station battery, or even an uninterruptible power supply (UPS) for complete isolation from the battery. If the substation automation system is the sole means for operating the substation, it might be desirable to have a backup or standby system. However, very few utilities install dual RTUs for reliability; and most IEDs include a user interface that can be used in an emergency.

Because of the data volume, a utility should try to stay as close as possible to the selected vendor's standard interface design. Requiring a substantial redesign just to avoid training personnel on how to use the new design is seldom economical. Likewise, a utility should try to use one of the more common "standard" protocols wherever possible. If the currently used protocol is not of fairly recent design, it is very possible that the protocol will not be able to support the functions that the modern IED can support. A significant capability is that of remote relay testing and changing of settings. This can save significant labor time and expense by avoiding trips to the substation. It also provides the opportunity for frequent spot-checking to ensure overall protection system performance and reliability. Some IEDs support multiple sets of protection settings that can be activated remotely, or changed dynamically, as conditions change on the power system.

5. SUBSTATION INTERFACE AND SUBSTATION AUTOMATION SYSTEMS

Substation automation activities tend to divide into two fairly distinct approaches. For convenience, these are defined here as either a *substation interface system* or a *substation automation system*.

The substation interface system (SIS) is primarily a pass-through system that can use a PC or other small computer for the user interface which, in turn, is typically supported by a very small local database. A generalized block diagram of the SIS is illustrated in Figure 9-4.

A "device interface unit" (DIU) is used to connect each IED to the substation LAN. Usually, the DIU is a protocol translator, but it may also perform other functions to achieve LAN compatibility. Some IEDs will require a DIU to be used with each IED, whereas others can use a common DIU for several IEDs. In most implementations of the SIS, two LANs are used. The primary LAN is dedicated to real-time data and control functions, and the second LAN is intended for accessing IED functions, such as settings, event records, historical data, and so forth.

The user interface displays current real-time data and such historical data as can be obtained from the IED storage registers. Limited alarming capability is frequently

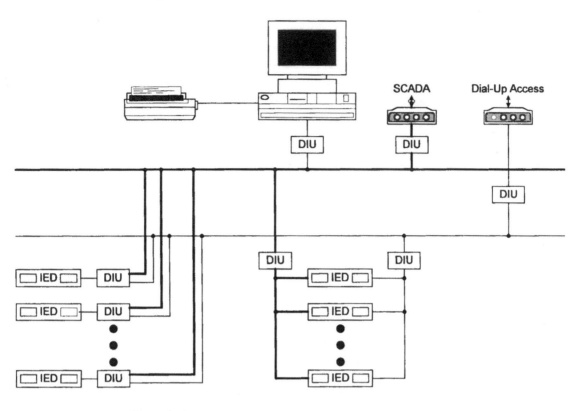

Figure 9-4 Generalized diagram of a substation interface system.

provided. The primary function of the SIS is to provide protocol translation so that one or more SCADA master stations can access real-time data from the substation. Some SIS provide pass-through access to the IEDs in the substation. That is, users wishing to obtain non–real-time data from the IED must use the IED manufacturer's programs for such access. Thus, if the IED is an ABB relay, ABB's ECP program would have to be used. If the IED is a GE relay, it would be necessary to use GE's equivalent to the ECP program. The net result is that the user must have multiple programs, each with their own user interface look and feel, to access all the IEDs in the substation. The lack of a common look and feel can impose significant training and operating problems.

Most substation interface systems do not support local automation functions by themselves. Thus, restoration schemes, are supplementary to, and not supported by, the interface system; and may not even be provided by the SIS vendor. In some cases, the configuration of the SIS system is such that timing constraints in the transmission and availability of common data make it impractical to include many automation functions.

The substation automation system (SAS) typically provides all the functions of an SIS and has the capability for many more. Figure 9-5 illustrates a block diagram of an SAS.

Usually, the SAS is designed around a database machine; that is, it has a comprehensive database that contains, at least, all the real-time data available from each IED. Often, a historical database is supported that allows the user to define what data is to be retained for plots and trend listings. Because the database concept supports access to non–real-time IED data, there is a more common look and feel for working with this data. Thus, relay settings, fault records, operations records, and such, all are accessed by the same sequence of user steps, and keystrokes always have the same meaning. If desired, password security can be made common throughout the system so that individual passwords are not required on a per-IED basis.

A major distinction between the SIS and the SAS is the support for supplementary automation functions. The SAS is normally based on a the use of a high-speed LAN so that data is exchanged and events take place in a real-time environment. The SAS computer, a PLC, or a combination thereof, can be used to accomplish automatic reclosing and restoration sequences within the substation. Restoration sequences are normally much more complex and are based on current-operating conditions because of the capabilities

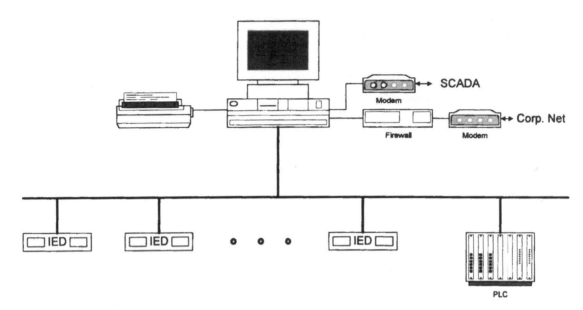

Figure 9-5 Generalized block diagram of a substation automation system.

of the SAS. Likewise, the SAS will normally have the ability to support distribution automation functions, such as feeder and load transfers within the same substation, and by using peer-to-peer communications capabilities, automate and integrate the transfer of feeders and loads between substations. A significant advantage of the SAS is the elimination of a need for a second LAN to perform non–real-time functions.

6.   CANDIDATES FOR AUTOMATION

Once an SAS is installed, ultimately there will be no real limit to the substation functions that can be automated. Some of the more obvious functions that can be implemented with today's capabilities include the following:

*   Alarm functions
    *   Transmit one alarm message, send details on request
    *   Apply filters to nuisance alarms (low air, momentary communications drop-out, and such)
*   Capacitor bank control (time, voltage, or both)
*   Dead line, dead bus reclosing
*   Voltage regulation (including intentional reductions)
*   Transformer bank paralleling
*   Load shedding
*   Underfrequency load rotation
*   Automated relay carrier testing

Alarm filtering can greatly assist the SCADA system operator. Many SCADA system specifications include a re-

quirement that the system be capable of processing 2000 or more alarms per second for 5–10 min. This requirement is used as a method to ensure that the SCADA system has enough computer power to ride through the most severe power system disturbance without bogging down or crashing. However, in reality, there is no system operator or dispatcher that can reasonably respond to 2000 alarms per second. Hence, anything that can be done to consolidate alarms to focus attention on the specific event that triggered the alarms, or that will filter out nonalarms, is a major asset to the utility and the system operators. Events that become alarms only if they persist for a period of time are the easiest to filter—just include a short time delay in reporting so that if the event returns to normal within the time delay period, it is not reported as an alarm at all. However, it may be desirable to include the event in an events log for historical analysis or for equipment condition monitoring.

7.   THE SUBSTATION AUTOMATION SYSTEM AS A CORPORATE DATA SOURCE

The combination of a substation automation system and communications technology results in a new corporate data source. In the past, substation data appeared to be the exclusive province of the operations (SCADA) department or the protection department of the utility. Other departments had to make requests for data and then wait for the requests to be fulfilled.

Today, the problem is not so much data archiving as it is data retrieval. Without good retrieval tools, data archiving becomes a "write-only" process. That is, the user knows the data is available somewhere, but does not have a means to easily access the data. Along with data accessibility is the issue of data security.

In the past, an individual had to have a company identification card to get in the front door of the control center building, a card key or special key to get into the control center area of the building, possibly a second card key or high-level access code to enter the control room itself, and finally a login password to get access to a console in the control room. When the Internet, an Intranet, or simply widely available computer terminals are used, the physical access barriers are completely bypassed and the only security is the login password. It is generally accepted fact that without vigorous enforcement of password change and control, any security is, at best, limited and, probably, nonexistent. In the United States, a 1997 report by the National Security Telecommunications Advisory Committee stated:

Utilities Vulnerable to Hackers

> Competition is forcing utilities to increase their links to suppliers, customers, and one another . . . and to shift to standard protocols. All those changes add to the risk of attacks. . . . Utility substations have the highest IT security vulnerability in the power grid because many of the automated devices used to monitor and control substation equipment are poorly protected against intrusion.

Many data-archiving techniques in use today are based on relational databases, and employ standard query language (SQL) to extract specific data. The volume of available data requires the application of new concepts and methods, sometimes referred to as "data warehousing." Essentially, this means that all data is easily available to anyone with the proper authorization. Note, however, that some care must be employed in providing access to the data warehouse. It is easy to generate a "query" to a database that will bring even the larger servers to a virtual halt.

Almost all data can be "public" in that anyone in the utility organization should have access. Some data can be "public–public" in that customers and others outside the utility can have access. In fact, the open access, same-time information system (OASIS) and Real-time information network (RIN) requirements that are being established in the United States mandate that the information be available to anyone who wants it, virtually without restriction. Other data and control functions (such as breaker operate commands, relay protection settings, and such) require strict control and highly secure access.

The use of specialized data retrieval nodes can provide a certain degree of security. Highly secure nodes may transmit an authorization or authentication message as part of every transaction. Such messages can be dynamic, in that they change in some way each transmission, or can be strongly encrypted. Personnel without access to such specially designed hardware are thereby essentially locked out.

Data mining is the latest "buzzword" that is used to refer to building massive databases and then applying various programs and techniques to search for new or undiscovered patterns that might offer something related to a competitive advantage, selling opportunity, reliability improvement, or undiscovered weakness in the power system design.

8. UNANTICIPATED CONSEQUENCES OF SUBSTATION AUTOMATION

There are several unanticipated consequences of substation automation and the use of IEDs in the substation. Basically, all of them are related to data, its availability, and its speed of transmission. This is best seen by considering the typical "RTU-monitored" circuit breaker position and the typical "IED-equipped" circuit breaker position. In the RTU configuration, typically the per phase currents and voltages are monitored, along with a three-phase total of watts and vars. There may be one or two control outputs and one or two corresponding status inputs; for a grand total of about 10 analog and status points per line or transformer. A typical IED, by contrast, can provide as much as 400 real-time analog and status points. The impact of this is discussed in greater detail in a later section of this chapter. What follows is a brief summary.

The first impact of this data explosion is on the communications system. If a utility is able to meet its required scan times using RTUs operating over 1200-bps communications lines, then the additional data will require data communications at about 48 kbps. Add the opportunity to gather non–real-time data on an as-needed basis, and the nominal communications requirement seems to be about 56 kbps. Certainly, the communications technology exists (in fact, it is widely available at moderate cost), but the cost to a utility to replace and upgrade its entire communications infrastructure to the higher communications speed will be very substantial.

Fortunately or unfortunately, the higher communications speeds are not an urgent requirement because most SCADA systems in use, or even being specified today, are incapable of receiving, processing, storing, using, and displaying the massive volume of data that is available. Brute-force upgrades to bigger databases, bigger and faster

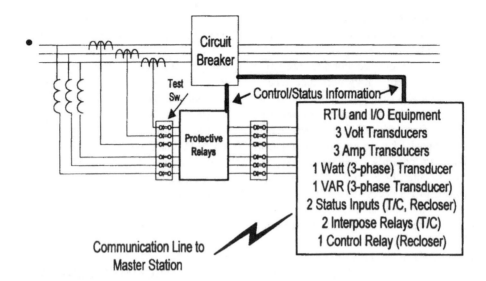

Figure 9-6 Connections for SCADA RTU checkout.

processors, bigger and faster disk farms are not a solution; although they may be prerequisites to a solution.

The traditional SCADA user interface that requires an operator to acknowledge and define the disposition of each and every alarm (limit alarms, transitions from normal to varying degrees of abnormal and return, and so forth) are just not possible. Operators dedicated 100% to such a function would not be able to keep up in even mild system disturbances. Clearly, new technologies in alarm and event processing are required. Alarm identification, cause-and-effect analysis, filtering, reliable, and virtually fail-safe sorting of alarms into various criticality classes, all are urgent needs.

In a like manner, the programs we like to call "advanced applications," such as optimal power flow, state estimation, and contingency analysis, will require whole-sale change and redesign. At the transmission level, virtually all such programs use a single-phase analysis, on the assumption that all three phases are substantially balanced. There are more and more indications that this assumption may not be universally valid. Moving from single-phase representation to three-phase representation of the power system, recognizing that the per-phase quantities are not truly independent, requires new and perhaps undeveloped techniques of analysis.

It is a well-known fact that an electric power system is mathematically "nonlinear." Because of the complexity of nonlinear analysis techniques, computational convenience (and necessity) has resulted in the application of approximation techniques to develop "linearized" models so that analytical methods can be applied. However, such linear-

ized models are usually based on the assumption of "small" deviations from a nominal operating point. Current practice is to operate existing power system facilities over a much wider dynamic range to maximize asset utilization. It is essential, in these circumstances, to frequently review the basic assumptions that have been made to ensure that they remain valid. Failure to do so can easily give a false sense of security.

On the positive side, the huge volume of additional information should make it possible to achieve much greater consistency and reliability in the calculated results. Application of advanced statistical techniques may actually enable the final result to have a higher degree of accuracy than any individual piece of data going into the calculation. The question then becomes one of how can there be assurance that the data coming from the IEDs in the substation are being properly received, processed, stored, and displayed at the SCADA master station or the corporate data warehouse.

Today, this assurance is obtained by performing a point-to-point checkout of every point in the RTU back to the SCADA master station. Two technicians, one at the substation and one at the master station console, work together to verify that a specific input to the RTU analog or status input card shows up in the right location on the SCADA display, in the alarm printout, and in most other locations where it is supposed to appear. For analog points, the RTU input quantity is known (i.e., 100 A) and the technician at the master station verifies that "100 A" is displayed. The substation connections required are illustrated in Figure 9-6.

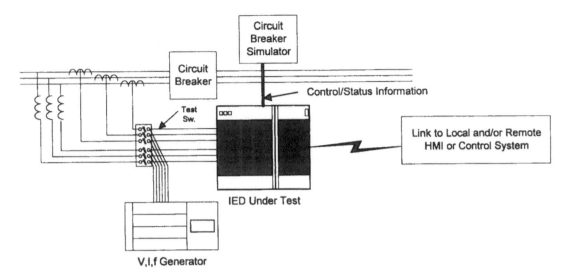

Figure 9-7 Connections for point-to-point testing an IED.

Typically, the checkout requires that the substation technician generate a dc test current in the range of 0 to ± 1 mA and apply same to the analog input card. For status inputs, all that is required is to ground or open-circuit the corresponding input point. Because the RTU and the protective relay are normally separate units, independently wired, it is usually possible to check out the RTU without disabling the protection. A medium-sized substation RTU can be checked out completely in about 1 day.

By using the 40-multiplier developed in the foregoing, we roughly conclude that an IED-equipped substation will require about 40 days to check out. In reality, the time may well be even longer because of the amount of equipment that must be used to check out an IED. First the test switches must be opened so that predetermined currents and voltages can be injected into the IED. The typical test box that is used to generate the precision voltages and currents costs about $40,000 and requires an experienced operator to manipulate the dials and controls to achieve the desired test inputs. It is even more difficult to achieve the desired objective of only one variable changing at a time to eliminate any ambiguity between cause and effect. A test setup is diagrammed in Figure 9-7, and pictured in Figure 9-8.

As Figure 9-8 indicates, point-to-point testing of an IED is not an easy task. Fortunately, a new development in IED design will make point-to-point testing much easier. In its simplest form, the concept, known as SCADA-REDI, allows the test technician to inject a specific data value into the communications registers of the IED. The specific data value is then transmitted to the SCADA master in the normal manner, where it is processed and displayed. Because the data injection device is a laptop computer connected to the RS-232 port of the IED, all the temporary cables and expensive test equipment are eliminated. Furthermore, the IED continues to perform its normal protection functions. Only the communications registers of the IED are affected. Because the data injection process can also be done remotely by use of a modem, the

Figure 9-8 Photograph of actual test configuration for one IED.

potential exists to eliminate the need for a technician in the substation. The entire process can be controlled by the technician at the SCADA master. Because the injection device is a laptop computer, part of the process can be automated, such that the laptop computer steps through a set of predefined values as quickly as desired by the SCADA technician. Finally, at some point SCADA systems will incorporate the necessary software to allow complete automation of the process, such that a substation can be completely checked out in a matter of minutes. Figure 9-9 illustrates the substation setup.

A final comment relates to data transmission speeds and the use of LANs in the substation in place of direct wiring of controls and indications. With direct-wired panels in the substation the operation of the "trip" or "close" switch results in a virtually instantaneous change in relevant indications. Lights change from one color to another, and all the needles on the ammeters, wattmeters, and voltmeters quickly swing to their new values. Even transient conditions, such as cold-load pickup amperes, can be observed and perhaps the magnitude estimated from the swing of the ammeters. With a computer-based user interface, and data transmission by a LAN, there can be a significant time lapse (several seconds or more) between the issuance of a command and the observation of the results. Usually, the brief transient indications are neither captured nor observed. Typically, a maximum update time on the order of

1 s is allowed between the time that a command is issued and the results appear on the user interface screen. Anything longer can create a bad impression and result in adverse comments or positive dislike for the substation automation system by the substation operating personnel.

9. COMMUNICATIONS WITHIN THE SUBSTATION

This section covers developments in communication systems as they apply to protective relays and power system information and automation networks. Over the past decade, advances in the microprocessor have permitted new and innovative ways to transfer information between applications. This is evident with the growth and acceptance of communication networks, such as the Internet, as a means of transferring or accessing information from around the world. As corporate and department managers learn of this available information from various sections of the power system, the need to easily access this data will be a requirement of the communication's network and its components.

With the increasing demand for power automation, a need arises to develop a system communication strategy that can handle high rates of data transfer. The ideal solution is to have the ability for data collection from various

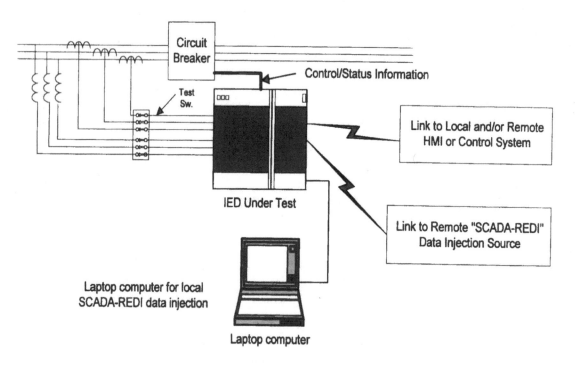

Figure 9-9 Connections to an IED for SCADA-REDI point-to-point checkout.

types of IEDs on a distributed network as well as a means to reliably control the substation equipment. Some high-speed communication interfaces embedded within the protective relay, which can also be called an IED, provide a means to communicate at rates of 1 million bits/s (1 Mbps) and higher. This type of system connects devices such as a programmable logic controller (PLC) and various types of IEDs, allowing plug-and-play automation solutions. A plug-and-play solution refers to a system in which components can easily be set up with minimal efforts and in some cases, the only setup required is to configure a device's network address and make the physical connection to the network.

10. BACKGROUND OF THE TRADITIONAL PROTECTIVE RELAY

A protective relay is a device that monitors the power system for abnormal conditions and takes preventative actions to limit power system stress and to reduce equipment damage. With the early 8-bit microprocessor-based relay systems, protection of the power system was the primary function, but advances in the computing power of a microprocessor permitted additional functionality, such as metering, breaker reclosing, and communications. The communication's interface in the typical microprocessor-based protective relay used EIA-RS232 serial communication port(s) connected to a VT100 terminal, or to a computer with a terminal emulation program. The protocol used was standard 8-bit ASCII; for which every number, letter or punctuation mark is represented by a set of 8 bits (1 byte). The EIA-RS232 interface provides a communications path between a single "master" and a single "slave."

Although these 8-bit devices had the capability to communicate at rates of 19,200 bps, the actual data throughput was much slower. This was because the device's protection function is assigned a higher priority than the communication process. In addition, the transmission of a single status indication (is a particular status set or reset?) required the transmission of 8 data bits (the ASCII character for T or F, or T or C, or some similar representation). In extreme cases, this inefficiency was made even worse by transmitting a complete word ("trip" or "close") instead of a single character. In any case, the priority assigned to communications processes required control of either hardware or software data flow between the master and slave devices to ensure that the information was received and processed correctly. An example of hardware data flow control, also known as hardware handshaking, is when a master "requests to send" (RTS) data to the slave and the

slave grants the master access through a "clear to send" (CTS) signal. When the slave device is not able to process any more data, it deasserts the CTS signal, thereby stopping the master from additional data flow. Once the slave is able to process more data, it asserts the CTS signal, thereby allowing the master to continue the data transfer. This stop-and-go process continues until the transaction is complete. This handshaking process also applies for communications in the reverse direction (i.e., for a slave's response back to the master).

Another type of common communication's interface for microprocessor-based relays is an EIA-RS485, supporting communication's rates of up to 19.2 kbaud. This serial communication technology allows a single master to communicate with multiple slave devices using a two-wire (half-duplex) differential multidrop physical layer. Each device, whether it is a master or slave, must have a unique address, and this address must be embedded in the master's command sequence. This command sequence is received, and the address is decoded by every slave device connected to the network. When the decoded address matches the slave's address, this device processes the master's request and replies with the response. Because of the "single master—multiple slave" configuration, applications using the two-wire differential pair to both transmit and receive data cannot make use of hardware data flow control. The net result is that using EIA-RS485 techniques requires greater processing capabilities in both the slave and master devices to ensure complete data transfer, while at the same time, not interfering with the protective relay functions of the slaves.

10.1 Peer-to-Peer and Multiple Master Communications

Typical relay communication networks, until recently, were capable of being configured only as a single-master device communicating to one or more slave devices. A recent advancement in IED communications has been the application of high-speed communication networks, allowing multiple masters and peer-to-peer networking schemes. Unlike the EIA-RS485 multidrop network, multiple masters can issue commands to several peers, and the responses from the peers can occur simultaneously. This requires a network architecture that prevents and avoids data collisions from multiple devices transmitting at the same instant of time. Or, the network architecture must include a technique to detect when simultaneous transmissions do occur and a method for retransmitting the data.

A major advantage of the multiple master environment is the distribution of the communication tasks to the various masters. These multiple masters permit the segregation of tasks during critical periods when events, such as faults or frequency disturbances, occur. An application of the multiple master is when precise control logic requires the timely supervision of networked IEDs, while at the same time, the system operators need information on the events. In a single-master environment, less important data collection has the potential to cause congestion on the network, not permitting the time-critical information exchange between the master and slave devices.

10.2 Token Ring Network

One technique commonly used in network communications is a token-passing architecture (IEEE 803.x). The token is a virtual privilege to network access and is passed between every device and node connected to the local network. For a device to communicate on the network, it must possess the token before any network services (read, write, and so forth) are permitted. Another important feature in the token-passing architecture is the internal policing of the token allocation time. Each device is allowed token possession for maximum amount of time before forfeiture of the token to ensure that every node has equal access to the network. In some token ring networks, if a token owner does not require the maximum time, the node relinquishes the token and remaining time to the next device in the token ring. After all devices in the network possess the token, the first device regains the token, and this is how the term token is derived. In addition, the amount of time required to make one complete loop in the token ring is called the token rotation time. There are other token-passing architectures that will hold the token for an allocated time, regardless of whether the node requires network access.

Token ring networks provide various techniques to exchange data from device-to-device on the network. As with the traditional EIA-RS485 networks, a token-passing network also supports master-to-slave communications. One difference in this type of command is that the master must possess the token to issue the slave's command sequence. Once the master owns the token, it can issue a command to a single, or commands to multiple slave devices. The command sequence is received by all devices on the network, and the addressed slave device(s) pro-

cesses the request after it obtains token ownership. The major advantage of this type of architecture is that, while the slave is processing the master's request, the network remains active, permitting other masters (or the same master) to process additional slave commands. When using the EIA-RS485, the master must wait for the completion of the master–slave transaction before another slave device can be solicited.

This is important when large data throughput is required. For example, suppose a master requests 64 bytes (512 bits) of binary metering data from each of 20 slave devices. For this example, the average response-processing time of the IED is assumed to be 100 ms. (The actual response time will vary depending on the processing capability of the IED.) The Table 9-1 shows the results of a comparison between an EIA-RS485 (9600 bps) and a token passing (1 Mbps) network.

10.3 Collision-Sensing Multiple Access Networks

Collision-sensing multiple access (CSMA) networks (IEEE 802.x) are normally based on Ethernet technology. Such networks allow each device on the network to "listen," and if the network is not in use, transmit data to other devices or network nodes. There is a finite probability that two or more devices might start transmitting data simultaneously, having listened and found the network idle. In such cases, the devices detect the data "collision" and retransmit the data after a random delay.

Experience has established that an Ethernet network suffers few collisions, and it actually transmits data faster when the network data load is not more than about 20% of the theoretical capacity. Thus, for a 10-Mbps network, up to 2 Mbps of data can be transmitted very efficiently. However, as the volume of data increases, the probability of collision also increases, with the result that the network data transmission drastically decreases to the point of being totally blocked. The obvious solution is to move to a higher-speed Ethernet—thus, 100-Mbit and 1-Gbit Ethernets are already becoming common.

The clear difference between token ring and CSMA, then, is that token ring is "deterministic" in operation. That

Table 9-1 Comparison Between an EIA-RS485 and a Token Passing Network

Network	Data transmit time	IED response time	Total response time
EIA-RS485	20*512/9600 = 1067 ms	20 * 100 = 2000 ms	3067 ms
Token ring poll	20*512/1M = 10 ms	20*25 + 100 ms = 600 ms	610 ms
Token ring–global variables	20*512/1M = 10 ms	100 ms	110 ms

is, the time to transmit a piece of data from one device to another is known and is virtually unchangeable. If there are many nodes on the token ring network, a device that needs to transmit data has to wait until it receives the "token;" this delay can sometimes be undesirable. On the other hand, the CSMA network will usually transmit data faster, except when there is a lot of data, in which case the data may never get through. As indicated in the foregoing, the studies performed as part of the LAN initiative have led to the conclusion that, overall, high-speed CSMA networks are preferred for future use.

10.4 Unsolicited and Global Data Exchange

Another advantage of these networks is the transmission and collection of unsolicited user-definable data (global data). Global data is broadcast from one device to all other peers in the local network. Every peer device receives and maintains a local copy of this global data for every other node in the network. The ability to transmit and receive global data is dependent on the peer's application and how the IED was implemented. Global data exchange requires network access. Updates can occur once per token rotation or whenever required in CSMA networks, or at slower rates if the peer does not need to update its global data. These global database-update rates are dependent on the implementation and capability of the IED.

Global data can be used when commonly required information, such as analog power quantities and binary input–output status words, would require fast-update rates in the master. In a master–slave-type network, this data would be polled by using the master and then retransmitted by the master to the other devices on the network, using valuable network resources. Another advantage of using global data exchange is the ability to customize the number of data points broadcast, thus, minimizing the amount of time required for transmission. In the event that the global data is not required from an IED, the transmission of global data points can be turned off.

10.5 Implementation of the IED Network Interface

To achieve the many advantages of high-speed networks, several considerations are required for the successful implementation of the IED's network interface. The high data-transfer rate is only as good as the weakest link in the entire system. Furthermore, flexibility of the IED and the ability to provide a means of user customization allows an ease of integration.

Because multiple master networks allow simultaneous command processing, the peer device must have the abil-

ity to process more than one command sequence concurrently. If the peer device allows only single-command processing, a lower-priority command may prevent time-critical command from being processed. This can be easily demonstrated in a substation automation application when two masters request data or issue commands to a common peer device. In this application, the masters consist of a PLC that accesses the network for automatic control and a host computer responsible for data collection and reporting status information to the system operators. If the IED's network interface allowed only single-command processing and the peer was busy processing a response to the host computer, a PLC that was requested to trip a breaker would be blocked until the command path cleared. The ability to concurrently process commands allows the PLC trip command to be executed and not blocked.

Another important consideration is the IED's performance and its ability to execute requests efficiently. Certain requests issued to the IED may not require immediate processing and can be deferred to a lower-priority processing task. Responding to metering requests and collection of event reports are examples of the commands that, although important, can efficiently be processed in a lower-priority task. In contrast, the time-critical control logic of the PLC is required to be processed in the IED's high-priority processing task for which delays owing to lack of performance, could result in equipment damage in the power system. In this type of application, a worst-case–timing study should be performed on the entire system to ensure that the critical timing events are accomplished. Key factors in determining the worst-case timing are the token rotation time and the IED's processing time that is dependent on the ability to prioritize incoming requests. As previously discussed, the LAN initiative design criteria is heavily dependent on specified performance considerations.

Flexibility of the IED is another key factor in the adaptation to user's various applications. Global data definitions are typically restricted to a limited number of available data points. An example of inflexibility is the points being on a fixed configuration. The ability to customize and allow the user to select which points are made available to other peer devices through the use of global data, provides the necessary flexibility required for ease of automation and integration of the IED.

10.6 Application—A Fully Automated Power Restoration System

The full utilization of the capabilities of a substation automation system are best illustrated by some detailed examples. The following discussion describes an application of PLCs, IEDs, HMIs (human machine interfaces), and networks; along with some of the reasons for particular selections.

The key to fast, reliable, repeatable, substation restoration performance is a high-speed communication link between all components. The link used in this application is a 1-Mbps Modbus Plus token ring-type network. A token-passing communications network is predictable in its performance, given multiple fault and restoration scenarios. It is not only the speed of the network that allows for restoration reliability, but also the intense use of global data variables. As described earlier, global data variables are broadcast over a token ring network every time a network device receives the token (permission to talk). Every other device on the network keeps a local copy of all other devices global data. This is typically stored in random access memory (RAM) and is refreshed every token pass. This allows high-speed data transfer without the need to poll by any of the network masters, such as, the PLC or the SCADA system. Some communications networks do not support the use of global data. In this case, a "master" (the PLC, in this example) must constantly poll each device for tripping status, breaker state, and control bit positions. Anytime polling is required, network capacity suffers, especially in larger stations where many devices require polling. The reduced network capacity may be acceptable in smaller substations, if fast power system restoration is not required and data acquisition and control are kept to a minimum.

The SCADA systems typically are programmed to expect acknowledgments to commands or scans within a preprogrammed time interval. If a command or scan is not acknowledged within this preprogrammed time, a communication error or time-out will occur. Various bit-oriented communication protocols are typically used as a method of efficient data transport from a SCADA master station to a remote location. This permits the use of 1200-bps communication links (by modem), without sacrificing command and scan turnaround timing. An argument can be made that current modem technology allows communication speeds in excess of 28,800 bps, thus not requiring the use of efficient bit-oriented communications protocols. Although this is true, it is rare that a modem will reliably connect with a remote at 28,800 bps 100% of the time. Typically, both modems will negotiate an optimum baud rate that may be substantially less than the maximum modem rating. If the SCADA system were designed to operate at the 28,800-bps level, time-out or communications failures will occur at a lower bit rate. The cost of line conditioning beyond the standard voice-grade channel is also a consideration when requiring high-speed links.

Many substation automation systems consist of SCADA systems only. The following application deals with a "true" substation automation system in which, along with SCADA functionality, fast automated restoration of service occurs after multiple types of line, bus, transformer, and feeder faults.

Figure 9-10 illustrates a typical distribution substation that will serve as an example application of an SAS. IEDs are placed at every breaker from A to V and connected along with the PLC and SCADA system to the high-speed communications network. The IEDs are placed at every station circuit breaker to provide protection, control, and analog and digital data associated with the breaker on which it is applied. A station PLC is used to coordinate and supervise circuit breaker reclosing after a major outage, such as that caused by a bus fault. The application described in this text deals with this type of fault because it requires intense communication between PLC and IEDs. Multiple restoration schemes are programmed into the PLC. The PLC then decides what the best method of restoration is, based on input from all of the station IEDs. This section will deal only with the 66-kV line (A,B,C) and bus tie (D) circuit breakers. The 66 kV bus restoration scheme is only a very small part of the substation logic.

An IED is applied as a 66-kV bus differential relay. The line current transformers (ct's) and transformer bank highside ct's are connected in parallel so that anything other than through current will energize the instantaneous overcurrent element (50P). A programmable logical bit in the bus differential IED is assigned as "bus test enable" and is used as the SCADA or local interface to the PLC for bus restoration permission. Each line- and bus-tied IED also utilizes a "permission bit" allowing the PLC to close its associated circuit breaker. If this bit is a 1, bus testing is enabled. If the bit is a 0, bus testing is disabled. Refer to Figure 9-11 for an overall picture of the bus restoration logic. Because the PLC logic is extremely intense in its diagnostics of the overall fault and reclose scenarios, Figure 9-11 shows only the basic logical flow and does not account for all bus restoration logic conditions. The following text will define, in more detail, the restoration logic.

When a bus fault occurs, the bus differential IED will send an instantaneous unconditional trip command by hard wire to all line (A, B, C,) and bus tie (D) IEDs. Because the PLC continually scans its local global database for any breaker operations or trip conditions, it will see the differential trip status, along with the line and bus tie circuit breaker status. The IED will determine if any breaker-fail conditions exist. Once it is determined that all of the circuit breakers opened successfully the PLC will initiate the restoration logic. The PLC determines all of these conditions in approximately 28 ms (1 scan of 412 ladder logic networks). If any line circuit breaker fails to open, the fault will be cleared by the remote breaker. In this case the PLC

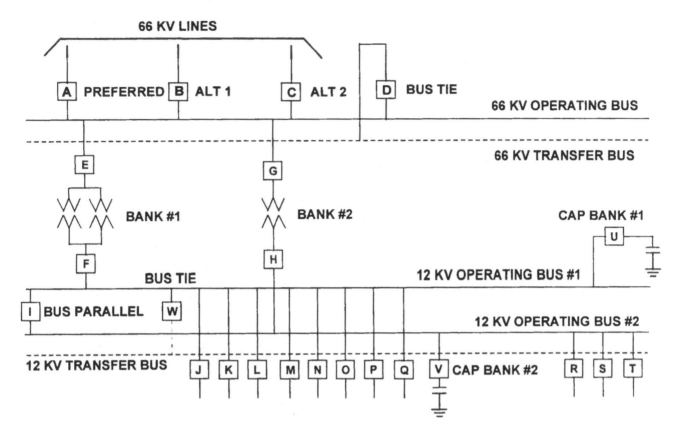

Figure 9-10 Sample substation.

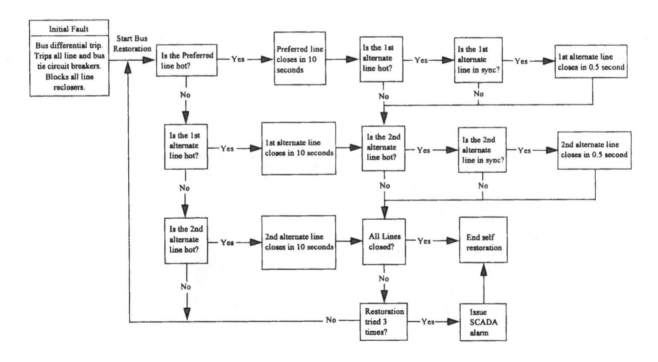

Figure 9-11 PLC logic flow.

will read a breaker-fail bit from the failed circuit breaker IED global data broadcast and lockout restoration. Each line IED also contains a PLC-supervised hot and dead line recloser.

After a bus fault occurs, a "recloser block" latch sets and disables the individual hot and dead line reclosers until bus restoration is complete. The bus testing starts 10 s after any line becomes hot, and all hot lines will test concurrently in a programmed sequence. Once a line circuit breaker closes and successfully energizes the bus, the PLC will check the status of the remaining lines. If they are hot and in syncronism, a close command is issued in 0.5 s. If any line is not hot or not in syncronism, the PLC will wait 60 s for these conditions to become true. If they do not, the PLC will abandon the effort and send an "unsuccessful restoration" alarm to SCADA. If restoration is successful, the logic will reset in 60 s. If the initial bus test fails, the logic will lockout.

The entire time that this logic is active and timing, SCADA scans are occurring because of the event. The high-speed network bandwidth combined with the use of global data variables assures that neither the PLC logical commands nor SCADA system are required to "wait" for the network. It can be concluded then that this type of system assures a very high percentage of timing accuracy and repeatability.

This type of restoration logic is not new and has been implemented in similar form for many years by use of electromechanical and solid-state relay systems. The major difference here is the greatly enhanced flexibility and reduced wiring costs. A PLC-based restoration system is almost infinitely programmable. To make a logic change with electromechanical or solid-state relays requires a protection system outage and many hours of rewiring and testing. Furthermore, control wires, sometimes in large

bundles, are run between cubicles to perform the same functions that can be accomplished using a single LAN connection.

10.7 Application—Data Collection Performance on a High-Speed Network

Another important use of the high-speed network is "real-time" data collection from the various IEDs in the network. The multidrop network shown in Figure 9-12 is a configuration of the sample substation in this application. The single-segment communication's network connects all feeder, line, transformer, bus, and capacitor control protective relays, along with the PLC and data collection master workstation. The master workstation maintains a system database of all IED metering, I/O status, alarm, and event information. As in the foregoing example, the protection IEDs have 22 global data points for acquisition of power system analog registers, bit-masked I/O status registers, and an IED event status register. Other less critical system information is polled by the master workstation on timed intervals or by detecting a change in the status register. Change in state of the status register indicates that a fault or an operation occurred in that specific IED, thus allowing the master to initiate polls to the IEDs.

The 25 devices (23 IEDs, 1 PLC, 1 master workstation) communicating in the 1-Mbps token-passing ring network has a nominal token rotation time of 25 ms for an average 1 ms/device. Each IED's global database updates the networked peer IEDs at a rate of 60–100 ms. Every peer device on the network will then maintain a completely updated database of all other peer's global data every six token passes, or 150 ms.

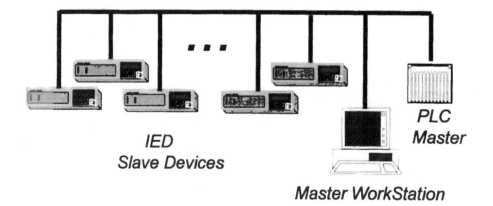

IED
Slave Devices

PLC
Master

Master WorkStation

Figure 9-12 Multidrop substation network.

In this application, the master workstation uses the global data, stored locally in the network interface, to create a substation database for all IEDs in the power system. Processing overhead in the master station for disk storage and MMI display limits the update rate of the power system database to less than 1 s. Typically, operators and technicians are accustomed to instantaneous electromechanical-metering responses and breaker state indicators. When using a virtual instrumentation panel, an update time of 3 s, or more, is usually unacceptable in the operator's perception questioning whether the breaker operated. Thus, the high-speed communication's network is important when the master station's MMI provides a means to issue commands, to perform various IED functions, such as tripping and closing breakers, and to acknowledge alarm conditions.

11. OTHER COMMUNICATIONS ISSUES

Communications within a substation can be accomplished easily and at a relatively modest cost when the appropriate provisions are made during construction. Typically, conduit or other physical provisions are installed from each current and future device location back to a central location. Then, as needed, twisted-pair wire, coaxial cable, or fiber-optic cable is installed.

The problem in existing construction is quite different. Most often there is no spare (or even any) conduit available to each device. As a result, it becomes necessary to pull back gravel and lay down some new conduit if physical connections are to be used. Alternatively, some utilities have installed conduit in the substation superstructure to avoid the burial costs.

Under these conditions, for existing substations, it appears that low-power, unlicensed spread-spectrum radio between devices in the substation and a central controller may be the most economical approach. Fortunately, low-power radio is becoming quite reliable and is capable of operation under most weather conditions.

12. COMMUNICATIONS BETWEEN SUBSTATIONS AND INFORMATION MANAGEMENT SYSTEMS (IMS)

This section describes the most popular communications technologies for master–remote communications. Most of the methods in use or being contemplated today are outlined, even though some may not be practical because of cost, noise, isolation licensing, or location considerations. Cost figures are based on 1998 dollars.

12.1 Leased Four-Wire Telephone Circuit

A leased line is typically the most reliable, and most costly, means of communications between a master station and a substation. Installation costs can be on the order of $1500 or so. Monthly recurring costs range from $50 to more than $250, depending on base cost plus a mileage factor.

Leased lines can be party-lined (that is, have more than one substation connected to the line), and usually will support data rates up to 28.8 kbps, although this can vary depending upon several factors. Reliability is in the range of 80–99%, and availability can approach 100%.

12.2 Dial-Up Telephone Circuit

The cost of a dial-up circuit is normally that of a standard business connection, approximately $40–$80/month. There may be a nominal installation charge for the circuit. An isolation device at the telephone demarcation point may be required for ground fault isolation.

When the connection is established, the dial-up line is normally as reliable as a leased line, as long as the connection is maintained. Availability is usually good, but is not guaranteed, especially during high-calling periods when an "all circuits busy" condition might exist. Unfortunately, high-calling periods seem to coincide with periods when substation information is needed most, such as major storms or other disasters.

Under some conditions a dial-up line can be multi-dropped, but this is not true for party-line operation. Once the connection is made, the dial-up line will typically support data rates up to 28.8 kbps.

12.3 Multiple Address System Radio (FCC Licensed)

Costs for multiple address system (MAS) radio are on the order of $2500 for each remote unit, and $12,000 for each master location. Installation could require the use of an antenna to achieve line-of-sight communications. Each site must be licensed by the Federal Communications Commission (FCC).

Reliability and availability of MAS radio can be equal to or better than that of a leased line. However, great care must be taken to ensure that a substation cannot "access" or respond to multiple master stations. Because radio is being used, the only reasonably guaranteed method of isolation is to make sure that all device addresses (that is, the RTU address) are distinct and not duplicated anywhere in the system. Although the FCC tries to ensure that frequency allocation includes adequate buffer distance between locations with identical frequencies, variable propagation and other characteristics of radio do not preclude the occasional "overlap." Hence, the need for dedicated addresses.

The FCC processes license applications in 90–180 days and, once granted, the MAS master and at least three remotes must be installed within 1 year or the license will be revoked. Essentially, all frequencies in this band have now been allocated; therefore, unless a utility already has frequency allocations, it is unlikely that a license can be obtained.

12.4 Unlicensed Radio, Spread-Spectrum

Unlicensed spread-spectrum radio is typically a low-power and short-range system. Therefore, to cover a wide area a large number of repeaters will be required. As a result, the infrastructure for this method of communication can be quite large; although once installed the cost per remote terminal unit is on the order of $500–$1000. Typically, reliability and availability tends to be very high (on the order of 99%) for a properly installed system. The most common data rate is 9600 bps, but the actual data throughput is determined by the number of intermediate repeaters that a message must traverse.

In some areas of the United States there are specialized companies that have built the infrastructure for unlicensed radio communications and make the service available at a fixed or variable charge per month. In some cases, the charges can approach that for a traditional leased line. However, for isolated locations, a third-party solution may not be available.

12.5 Cellular or PCS Systems

As the coverage of cellular or personal communications services (PCS) equipment increases, this may be an option to consider. The typical service has a fixed monthly charge, on the order of $25–$50; and an air time charge. The air time charge essentially eliminates the option of using "full-period" or real-time data. The remote location must be equipped to detect abnormal status changes or analog limit violations and initiate a call to the master station. (This mode of operation is typically called report by exception.) Obviously, the master station must be able to accept unsolicited data reports from the remote station; and sometimes the normal sequence for performing control functions will require modification.

12.6 Trunked Radio—Data on Voice

The trunked radio system is very expensive initially and generally includes a fairly large recurring operating cost.

Most often, a utility will have trunked radio to meet its needs for voice communications. If the system already exists, the incremental cost of adding data capabilities may be acceptable. However, the ability of the system to support both voice and data may be compromised during periods of high voice traffic, such as system problems. It may be necessary to impose priorities on the system to ensure that minimum data transmission requirements are met.

12.7 Satellite Techniques

Satellite communication is an emerging technology that is probably among the more expensive alternatives at the present time. However, developments are moving very fast, such that, in a few years, it may be the quasistandard technology.

There are various competing approaches being considered. Some use a few geostationary satellites, whereas others use large numbers of low- or medium-orbit satellites. Cost will certainly be a consideration. Today, it is possible to achieve reliable communications in just about all weather conditions. However, severe weather can cause "fade-outs" which is just about the same time that data is required from the substation. The geostationary orbit satellites do present a problem because of the noticeable delay (up to 0.3 s) in data transmission. Typically, this will impose a requirement to modify the control sequence in the SCADA master to allow enough time for the sequence to be completed.

12.8 Fiber-Optics

The fiber-optic technique is currently viewed as probably not required in the distribution substation application. The initial purchase of equipment, and the installation of fiber-optic cable is usually quite expensive. Installed cost of fiber cable can be up to $10,000/mile. Fiber transceivers costing about $1500 each are required at each end. Because of the relatively small amount of data that may have to move between the substation and the master station, fiber-optic techniques are generally seen as extreme overkill. However, if other factors create the opportunity for installation of fiber-optic communications, this will certainly provide ample bandwidth for any future data communications requirements.

13. SMALL SUBSTATION ISSUES

A substation in the traditional sense consists of at least one transformer and some switching equipment. The bare minimum would be an incoming line with a switch, a fused transformer, and a low-side circuit breaker or recloser. Depending on the location and application, a small substation might consist of a looped-through high-voltage line, a fused transformer, a low-side circuit breaker, and one or more feeder breakers. These substations are typically used to provide electric service to areas with low customer density. In many cases, low customer density areas are perceived as being "rural." Figure 9-13 is intended to depict a typical rural substation and its single-line configuration:

There are several practical problems associated with rural substations when it comes to improved quality of service. Typically, the substations were built over a long time period and have no provisions for the addition of SCADA capabilities. The distribution feeder lines out of the substation tend to be quite long, with a corresponding higher exposure to interruptions caused by nature (weather, animals, or other) or humans (hunters shooting at insulators, vehicle accidents, and such). When interruptions do occur, the first indication that there is a problem is usually when a customer calls in with a complaint. There follows a normal, but time-consuming, response that finally results in the dispatch of a repair crew to the presumed trouble spot. Typically, the crew first goes to the substation, which may require a significant travel time.

In summary, the perception, if not the reality, is that rural substations have more interruptions than a typical urban substation; and that the time to restore service is longer than for an urban location. Open access will require that greater emphasis be placed on customer service and electrical service reliability. Achieving improvements in these areas will be particularly dependent on obtaining real-time or quasi–real-time data from the substation;

which will require the addition of some SCADA functionality to the substation.

There is a large installed base of "small" substations, which typically consist of a radial or looped-through high-voltage source, a single transformer, perhaps a capacitor bank, and one or more feeders. Standard SCADA installation practices frequently result in cost estimates that make it difficult to justify the addition of SCADA to a small substation. There is a great need for cost-effective methods to add SCADA functionality to existing small substations, and to include SCADA functionality to new construction.

It is the usual experience that engineering, installation, and documentation costs for SCADA additions exceed, by far, the actual cost of the SCADA RTU. This paper discusses opportunities and techniques for achieving substantial savings in all cost areas. Even if "real-time" operational data is not required, the concepts presented in this chapter can be applied to allow retrieval of extensive data related to the operation of the substation, quality of service at the substation level, and equipment condition.

Achieving SCADA functionality in a substation implies two different communications issues. One is communications between the substation device and the remote terminal unit (RTU) or its equivalent. The other is communications between the RTU and the master station.

14. EQUIPMENT AND INSTALLATION CONSIDERATIONS

This section is primarily concerned with the addition of SCADA functionality to existing substations. Similar methods and techniques can be used for new construction. In fact, the problems associated with existing facilities, as

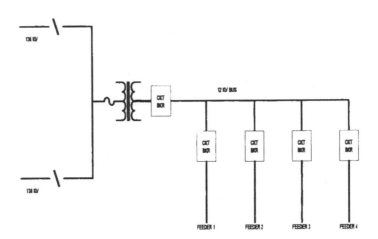

Figure 9-13 Small substation configuration.

outlined in this chapter, can be used to justify the installation of spare conduit and other provisions that will make future additions much easier.

14.1 Traditional RTU Equipment and Installation

Many small or rural substations are completely outdoor construction. There is no control house; hence, an RTU in an outdoor cabinet is required. The RTU can be powered from the station battery if one exists; or the RTU can operate from the station service source. If the station service is used, a UPS is strongly recommended to ensure SCADA functionality is available when the station power is unavailable. The UPS operate time should be at least twice the normal travel time to arrive at the substation, and a frequently of a fixed time of 4 or 8 h is used.

For outdoor construction, the protection equipment is typically located in a weather-proof box attached to or adjacent to the circuit breaker or recloser. On most feeders, only overcurrent protection is used; so it is quite likely that a three-phase voltage source is not available at the breaker position. If conduit is available, it may be possible to bring a bus voltage source to each breaker position, so that installation of watt and var transducers can be accomplished. Otherwise, the only analog values that can be captured will be the phase and possibly ground currents.

If a voltage input is available, it is possible to obtain self-powered volt, watt, and var transducers. However, this will increase the burden on the PT, so the effect must be evaluated. Some current transducers are self-powered from one of the current inputs, but the burden must again be checked. Otherwise, a separate power source must be provided for the transducer.

The traditional RTU solution requires that analog values be represented as a 0–1 mA dc signal. If the transducer is located at the breaker position, it will be necessary to provide shielded twisted-pair cabling back to the RTU location. The alternative to using a dc signal back to the RTU location is to mount the transducers in the RTU and then use heavier cable to bring the PT and CT leads back to the location. Typically this alternative is not used because of the cost of cable installation, or other costs.

In addition to the analog signals, the RTU will require some connection to the trip and close circuits of the breaker, as well as status contacts showing the open or closed status of the breaker. Either heavy-duty interposing relays will have to be mounted in the protection enclosure, or high-ampacity cables will have to be installed from the RTU location back to the breaker location.

Cost estimates to provide a traditional RTU approach to obtaining SCADA functionality in a small substation (see Fig. 9-13) range up to $75,000–$100,000 or more. These estimates do not include the communications media between the RTU and the SCADA master station. A typical breakdown of this cost estimate includes:

$20,000–$25,000: RTU, transducers, cables, conduit, miscellaneous hardware
$20,000–$25,000: Installation and commissioning labor
$25,000–$50,000: Engineering, design, and drawings

Some utilities have found that these costs tend to underestimate the actual work because the existing drawings have not been kept up to date and substantial additional work is required before the actual design work for the new additions can even start.

An alternative to the traditional self-contained RTU is the distributed RTU concept. In this approach, the I/O capabilities of the RTU are physically located inside or near the breaker protection enclosure, and the I/O equipment is then connected back to the station and communications controller portion of the RTU. The distributed I/O equipment will require a minimum of four wires from the central location to each I/O equipment: two wires for communications, and two wires for power from the RTU. In the absence of conduit, this approach is not much different than the approach outlined in the foregoing. If conduit does exist, there might be some modest savings because the communications and power wire for the I/O equipment can be of smaller gauge and, therefore, less costly and easier to install.

In conclusion, for most small substations, the traditional RTU cannot be cost-justified except in extremely special circumstances.

14.2 Use of Intelligent Electronic Devices

There are several definitions of *intelligent electronic device* (IED) available, the simplest being a device that has a communications (com) port that can send or receive data. Some typical examples include microprocessor-based relays, electronic meters, and even transducers that include a com port. In what follows, IED will be used as a shorthand notation for microprocessor-based relay. Other possible meanings will be obvious by the context.

Virtually all the desired real-time or quasi–real-time data that would be used by a SCADA system is available from the IED. Typically, this includes the following list:

Per-phase current magnitudes
Per-phase voltage magnitudes (if voltage inputs exist)
Per-phase or total kW (if voltage inputs exist)
Per-phase or total kvar (if voltage inputs exist)
Trip–close status of breaker or recloser

On–off status of automatic reclosing
Target information following an operation
Distance to fault (if voltage inputs exist)
Ability to execute trip–close commands to breaker
Ability to enable–disable automatic reclosing

Most IEDs can provide much more real-time and historical data, but the foregoing list roughly represents the minimum that might be wanted by a system operator to respond to problems.

14.3 Substation Controllers

In some applications in very small substations, it may be possible to have a single IED perform the protective functions and, also, through supplementary inputs to the IED, gather the minimum data believed necessary for SCADA functionality. In such cases, the IED may also include protocol emulation, such as DNP 3.0, that will permit direct communications to the SCADA master station.

Usually, however, it will be necessary or desirable to use a substation controller in a centralized location in the substation to consolidate the data and serve as the SCADA interface device.

14.3.1 Classic RTU Interface

In this application, a RTU is installed to serve as the communications and protocol processor. Typically it will include communications ports that interface to a link to the SCADA master on one side, and to the substation IEDs on the other side. Usually, but not always, the RTU interface does not include a user interface (UI) device, such as a CRT or other visual displays. Thus, any human interaction in the substation normally takes place directly on the end device or its IED.

14.3.2 Programmable Logic Controller Interface

The design of this interface utilizes a PLC to perform the protocol translations between the SCADA master and the substation IEDs. Fundamentally, it represents an approach similar to the classic RTU in terms of functionality and operation.

14.3.3 Computer-Based Interface

This interface makes use of the capabilities of computer technology (usually a "PC" or a "workstation") to perform the protocol translations and other functions required in the SCADA interface. The most popular choice is the PC

because of its current lower cost, and because of its status as a "standard" device. Quite often a CRT and keyboard is provided to provide a UI of varying complexity. At a minimum, a single-line diagram of the substation is provided, along with a display of significant metered and status values. The ability to send control commands to the substation IEDs is usually included.

There are several issues that must be considered in applying a PC to a substation automation installation. Usually the PC is not made a critical component of the substation's normal operation, so that a failure will not impair the protection or other aspects of the substation. In other words, the net effect of a PC failure is the same as that of a failure in the communications link, or loss of RTU functionality.

Selection of the appropriate operating system for the PC interface can be very controversial. One camp is strongly of the opinion that Microsoft Windows technology is the best approach because of the versatility that can be achieved, and because there are a number of "off-the-shelf" applications programs that can be used to develop the UI and some communications functions. The other camp contends that, although Microsoft technology is excellent in an office environment, it is not well suited for real-time applications. Users have adjusted to the inconvenience of operating system and application software failures that occur in the office environment. Most of the time a warm reboot ("control–alt–delete") is all that is necessary to resume operations, although, occasionally, a complete power-off operation is necessary to get things going. However, when this type of problem occurs in a rural substation, located many miles from the nearest human support, the recurring cost of driving several hours just to reboot the computer system can be intolerable. Consequently, operating systems that are specifically designed for real-time operations, or those that have been in use for extended time periods (and, therefore, largely "debugged"), such as UNIX or LINUX, tend to be the first choice.

Another consideration is the "do-it-yourself" approach to substation controller design versus the purchase of a factory-assembled controller. Typically the do-it-yourself approach, using off-the-shelf software packages, is perceived as being the most economical. However, in such cases, the utility must ensure (enforce?) excellent documentation and records so that personnel changes, such as promotions or turnover, do not result in the loss of critical knowledge.

The factory-assembled controller is not without risk, however. There are a number of firms offering equipment, and there is no guarantee that all will guarantee long-term support or even remain in business. In this area, however, the utility industry can select from several well-established companies with a tradition of long-term utility support. Even so, insistence on quality documentation is a must.

Program backup is an essential function in all cases where general-purpose computer technology is used. The variety of backup devices is extensive, and ranges from magnetic tape to recordable compact disk drivers. Care must be taken to ensure that the backup technology remains viable. With the speed that technology changes today, backup devices can become obsolete or unusable. Do not forget that some of the first PCs that came out used 8-in. floppy disks as the media. (The Radio Shack TRS-80, for example). Today, neither the media nor the drive is available, so what might have been perceived as excellent backup practice at the time is essentially useless. All backup practices (and backed-up software) must be reviewed at least annually to ensure that the practices are still valid, and that the backed-up software can still be recovered.

Finally, it is probably impossible to overestimate the effort required to develop and configure a protocol emulation interface for communications with the SCADA master station. At the present time the DNP 3.0 protocol seems to be the most preferred and is approaching "de facto standard" status. Even so, building the database definitions is still a time-consuming task even when "DNP protocol emulation" is claimed.

14.3.4 Installation Considerations

The IED can offer an attractive alternative to the typical approach of adding transducers to existing electromechanical (EM) relay installations. One significant justification for doing so is that EM relays are rapidly becoming obsolete and replacement parts or new relays will become more costly in the future. At some point in time, the few remaining U.S. manufacturers of EM relays will announce that they are discontinuing the EM relay product line, or increasing prices to offset the increased production costs owing to low volume.

Second, an IED with the same or substantially greater functionality than the existing EM relays will typically fit into the same physical space; or actually require less space. This feature tends to simplify the retrofit process.

Finally, and perhaps most important, since most IEDs include a "generic" communications port, it is possible to use whatever communications method is the most economical in the given substation.

An interesting technique proposed for retrofitting IEDs into an existing substation is to mount the IED and its auxiliary equipment (such as a small battery power supply and charger, and a short-range radio) in a separate weather-

proof box that can be mounted to the side of the existing protection enclosure. The case terminals of the existing EM relays are then used as the connecting points for the wiring to the IED (Fig. 9-14). This approach offers substantial installation savings, as well as engineering and drawing costs as discussed in the following section.

Unless conduit from a central location to each breaker position already exists, it is almost guaranteed that the use of short-range radio communications techniques will be more economical than hard-wire or fiber optics. A secondary advantage is that the substation communications controller can be located in a convenient location, such as to simplify access to the station battery box, or to a structure that can support an antenna for communications to the master station.

There are several items to consider in installing the central communications controller. If the traditional RTU approach is used, typically the only requirement is to use a weatherproof enclosure. Most of the time the RTU will have an operating temperature range that does not require the use of supplementary heating or cooling. If a station battery is not available for RTU power, sealed batteries should be used if they are to be located within the RTU enclosure. The same comments substantially apply if a PLC device is used to act as the protocol converter. In the case of either the RTU or the PLC, the options regarding a user interface are normally restricted to some form of a digital-indicating device or an annunciator panel. Very few RTUs or PLCs are equipped to drive a CRT, which can provide a more detailed UI.

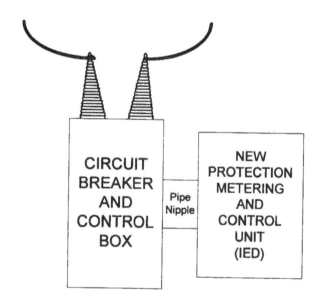

Figure 9-14 Connections of new IEDs to existing breakers.

Many times there is a desire to include other alarms and analog values in the information being sent to the SCADA master. This might include station battery voltage, status of disconnect switches, ground relay cut-outs, or others. Some IEDs will support a limited number of additional inputs and outputs as being user-defined. Most of the time these user-defined inputs and outputs are limited to status and control (i.e., digital) quantities. If the IED does not support additional I/O, it would be necessary to use an auxiliary device, such as a PLC, that accepts analog and digital inputs for transmission to the central controller. Or, because the PLC adds more cost to the installation, one must either select an IED that does support additional I/O or decide to forego the additional data.

It may be necessary to use supplementary heating or cooling in some environments when PC technology is used. "Hardened" PCS are available, but at a significant cost increase over the standard device.

It must not be forgotten that the most common time a user interface is needed in the substation is during bad weather when there are equipment problems. Some form of protective shelter for the CRT or other UI is, therefore, highly desirable. This comment is also applicable to the individual IED weatherproof boxes that are attached to the breaker protection box.

When we take into account all of the foregoing discussion, a small substation equipped with IEDs and short-range radio communications facilities would have a single-line diagram as indicated in Figure 9-15.

With use of the techniques discussed in the foregoing,

it is estimated that the final installed cost of a complete SCADA interface in a small substation, including a DNP 3.0 interface (but not the modem or communications path to the master station) can be reduced to about $35,000–$40,000. This cost is believed to be well within acceptable limits and can be recovered in a short time through more efficient operation, better crew dispatch and scheduling, and improved customer service and reliability.

In summary, there are many parameters that go into selecting and installing SCADA functionality in a small substation. By careful planning, standardization, and giving field installation personnel the freedom to do what is required, it should be possible to reduce the installed cost from about $100,000 to $35,000 or less. At the same time, the protection equipment is upgraded by replacing electromechanical relays with modern, microprocessor-based relays.

15. SUBSTATION AUTOMATION AND THE SCADA/EMS/DMS/IMS

Deregulation of the utility industry is creating a situation in which utilities must automate and acquire more information to remain competitive. Reduced costs of installing, operating, and maintaining the utility plant is a major objective. There are significant and quantifiable cost savings that can be achieved through the use of substation automation. Somewhat less tangible objectives include improved efficiency, improved system reliability, improved power qual-

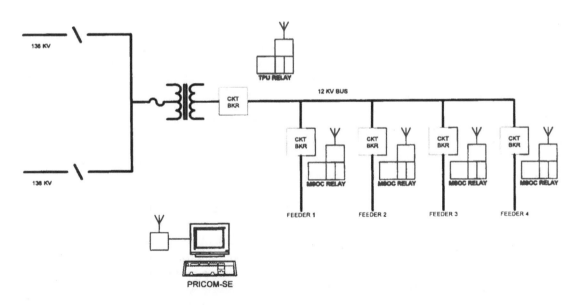

Figure 9-15 Small substation with short-range radio to achieve a "wireless network."

ity, and improved customer service. All of these items, in one or more ways, will be important in customer retention.

"Open access," sometimes referred to as wholesale and retail wheeling, creates a number of requirements for additional information about the power system. The objective of open access can be said to make maximum utilization of the capabilities of the electric power system, without violating any stability or equipment operation limits. Some states and the federal government are mandating the use of independent system operators (ISO) to control the operation and use of the power system in a safe and reliable manner. The ISO uses computers and complex software to perform real-time calculations of transmission capacity, stability, contingency identification and evaluation, and operating costs. All of these programs, however, are based on the use of power system data from the substation. It is self-evident that if the quality and quantity of information available is inadequate, then the results of the calculations will be less precise, thereby requiring greater "safety margins" to be applied. It can be expected that the ISO will demand more and better data to ensure that maximum benefits of open access are obtained, without creating undue risk of blackouts or other system-operating problems.

Operating settings of protective devices are typically established based on short circuit calculations made at various system load levels. It is impossible for these calculations to include all possible conditions on the power system; consequently compromises have to be made. The net result is that it is possible for situations to occur whereby relays will operate incorrectly.

Microprocessor-based relays typically permit multiple sets of settings, any of which can be put into effect based on system conditions or on direction of the system operator. There is no inherent reason why relay settings cannot be recalculated based on any significant change in power system conditions. These settings could then be automatically loaded into the relays as appropriate.

Sensors and transducers can be included in other substation apparatus, such that maintenance can be performed on an as-needed basis, rather than on a scheduled basis. In fact, it should be possible develop analytical procedures that would enable failure prediction and remedial action before the actual failure.

16. SUBSTATION AUTOMATION AND DATA VOLUME

Figure 9-16 illustrates a typical connection diagram for a circuit breaker with relays and transducers (and meters) to monitor voltage (V), current (I), real power (P), and reactive power (Q). The data that is sent to an RTU for trans-

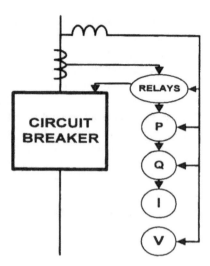

Figure 9-16 Typical connections with multiple devices.

mission to a SCADA master station would then include the following quantities:

3	Single-phase currents
1	Total three-phase watts
1	Total three-phase vars
3	Single-phase voltages (one set per bus)
1	Breaker trip–close status and control
1	Recloser status and control

The volume of data changes drastically when a microprocessor-based IED is used as the control element and data source. The connection diagram is very similar, as indicated in Figure 9-17. The major difference is that the interconnection wiring between transducers and meters is no longer required. However, the simplicity of the circuit is very misleading. Data available from the IED includes the following quantities:

Real-time values

3	Single-phase currents and phase angles
1	Neutral current and phase angle
3	Positive-, negative-, and zero-sequence currents and phase angles
3	Single-phase watts
3	Single-phase kWh
1	Total three-phase watts
1	Total three-phase kWh
3	Single-phase vars
3	Single-phase kVarh
1	Total three-phase vars
1	Total three-phase kVarh
3	Single-phase voltages and phase angles

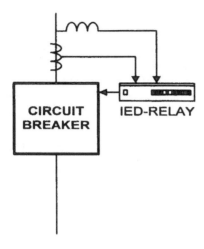

Figure 9-17 Typical connections with an IED.

1	Frequency
3	Single-phase power factors
1	Distance to fault

Demand values

3	Single-phase currents and angles
1	Neutral current and angle
4	Single-phase and total three-phase kW
4	Single-phase and total three-phase kvar

Status and control

1	Breaker trip–close status and control
1	Recloser status and control
25	Additional status points (minimum)
16	Additional control points (minimum)

In addition to the foregoing real-time data, there are sets of maximum and minimum values of currents (kW and kvar) consisting of magnitudes, dates and times. Finally, non–real-time data available includes up to 32 fault records, 128 operations records, 3,840 load profile records, and over 14,000 waveform samples.

Summarizing the situation, considering real-time values only, the following numbers would apply to a ten-feeder substation, not even counting incoming lines or transformers. It can be seen that, even ignoring all other data, the potential SCADA database is at least ten times larger when IEDs are used as the data source.

	RTU basis	IED basis
Analog quantities	53	430
Status quantities	20	270
Control quantities	20	180

16.1 Near-Term Effect of Data Volume

For the present, the presence of a SA system in the substation will enable the SCADA system to operate as though it is scanning a single RTU for data. However, the SCADA system database will have to be somewhere between 10 and 100 times larger than it is for the "pre-IED" world. This could be seen as suggesting that it will take 10–100 times longer to build a SCADA database (or require 10–100 times more people to do the work). Clearly, neither alternative is acceptable, so it will become necessary to use alternative means for database building and maintenance. It may be that the typical database architectures in use today cannot be continued into the future.

One potential approach is to take advantage of the uniformity of an IED. That is, the same model IEDs will have the same database structure. This can be used to build templates for each IED being used, and then replicating those templates. Another approach is to predefine standard objects, such as a single breaker with disconnects and bypass, along with all the data and control elements that apply. Then, the standard object is replicated as needed. It is necessary to keep in mind that there is no such thing as two identical substations, so object definition must be flexible. It is not advisable to try to build objects for all possible combinations, especially if there are only a limited number of opportunities to use the object. There will be a point at which the effort to define an object may be greater than just hand-building the appropriate database.

It is essential to pay close attention to naming conventions when defining objects. Automating the database-build process can impose naming conventions that are different from those currently in use. All personnel involved need to understand and accept the new conventions. Also, in building predefined objects, there is a tendency to use default values for such things as limits. Unfortunately, once databases are built with default values, unless a concerted effort is made to include actual values, the defaults will stay in place.

Once a database is built, it must be maintained to keep up with the constantly changing power system. Making changes by hand is not practical, simply because of the hundreds of points associated with each new IED. Also, some technique will have to be developed to ensure that updating the database did not change anything that should not change. Current practice of saving changes for a periodic (say, monthly) update probably will not work, simply because of the size of the update. Likewise, current methods for verifying (checkout) the database will have to be modified.

A similar problem applies to the development of displays for the SCADA system. The traditional method of hand-building displays cannot be used. The sheer volume of data would require too many pages of display for each substation. Pop-up windows or pull-down menus will be necessary so that operators can see a minimum set of summary data, and then request additional data when necessary. Display building techniques will have to incorporate the same database naming convention to make it possible to build automatic linkers between the database and the display.

Another area of concern relates to performance issues. Will it require 10–100 times greater CPU power to maintain today's performance standards? Disk storage requirements will increase dramatically to accommodate the greater database size. Some even question whether it will be possible to maintain the current practice of having shadow disks to speed up failover. Front-end processors will have to be upgraded to handle more data per unit time if the traditional 2-s scan of status values is to be maintained. One possibility would be to redefine several status points as being demand-scan-only, rather than continuous scan. The same comments apply to the processing of analog values on a 10-s or so scan basis.

The massive increase in size of the database has implications relating to database integrity scans and failover issues as well. SCADA systems will most likely have to employ report-by-exception and dead-band techniques to control the volume of database processing required. Currently, the practice is to make an "integrity scan" of all points on a periodic basis to ensure that the entire database is current. This may have to be divided up into smaller groups of data, with the check made less often. In a similar manner, most SCADA systems employ an initializing scan of all data after a failover, to recreate the real-time database. Such a process would seriously prolong the time for a failover process to complete. Again, alternative methods will have to be developed.

Current SCADA systems employ deterministic alarm processing. That is, alarms are preprioritized, and all are processed and require attention by the dispatcher. A major power system disturbance could trigger hundreds of thousands of individual alarms and events; clearly beyond the capacity of any dispatcher to handle in a responsible manner. There have been several attempts to build expert or rule-based alarm-processing systems; but most suffer from agonizing slowness before the initiating event(s) are identified. Again, new techniques may be required.

In a similar manner, many SCADA systems use deterministic processes to perform supervisory control. With IEDs involved, the control timing will be a function of the IED as well as the automation system and the communications process used. The dispatcher will no longer have a "feel" for how long a control sequence should take to complete; and the SCADA system cannot afford the time to suspend all scan traffic while a control sequence is in process on a communications line. The time-honored practice of "check before operate" may appear to be implemented, but in all probability the actual sequence of a control command will not truly verify that the end device has been selected before the execute command is issued.

16.2   Long-Term Effect of Data Volume

When the LAN initiative techniques become widespread, the concept of a RTU may not apply. Every device in the substation will have its own Ethernet address; and the database of objects will "build itself." However, display practices and techniques will also have to be drastically revised to accommodate the new data structures. When multimegabit Ethernet messages are being processed, it would appear that much greater front-end–processing power will be required to maintain a semblance of today's scan rates.

Unfortunately, it will not be possible to just reach a point where an "old" SCADA system can be retired and a "new and up-to-date" SCADA system installed. There will be a gradual transition from the current RTU technology to the new LAN initiative technology; implying that any SCADA system will have to accommodate both technologies for the foreseeable future. It may be that whole new SCADA/EMS/DMS system architectures will be required if current performance requirements are to be maintained.

Assuming we can solve the database, display, and performance issues, there remains the question of what is to be done with the data. Most EMS application programs start with a *state estimation* to provide a base point. State estimator (SE) techniques will probably have to be modified if they are to make effective use of the additional data. No good answers exist today as to what will happen to solution times. Currently, most application programs use a single-phase system model, mostly on the assumption that the unbalance at the transmission levels of interest is insignificant. As we push to higher and higher limits, however, unbalance may become a significant factor, and it will become necessary to move to a three-phase model. Once again, whole new techniques will be required to develop and maintain an accurate three-phase model, and new computer processing and matrix manipulation techniques will probably be required to handle the increased model size. Once the SE program is converted to a three-phase system, applications programmers will have to convert and upgrade the remaining applications programs, such as operator load flow, optimal power flow, contingency selection and evaluation, and so forth.

16.3   It Is Not All Bad News

There has to be a purpose and objective behind all the effort to utilize the increased data. Typically, more data allows the application programs to be more accurate in their results. This is not only desirable, it may be essential to cope with the problems associated with open access and open wheeling. Utilities should be able to do a more precise job of contingency selection and evaluation. Contingency avoidance techniques can be optimized for minimum cost or risk. The contingency analysis program could include remedial action recommendations as part of the solution. New techniques may make real-time voltage and system stability calculations a reality.

Finally, the advent of IEDs makes the goal of corporate-wide access to data a reality. New methods for accessing this data appear to be moving toward the web browser techniques. The only serious problem is that of secure access. If access to the IED database is through an SA system, firewalls and other security techniques can be incorporated into the SA system to provide a good level of security.

ACKNOWLEDGMENTS

The material on high-speed communications in a substation is based on a paper by C. B. Adamson of Southern California Edison Company, and S. Kunsman and M. Kleman of ABB Power Automation & Protection Division.

10

Protective Relay Digital Fault Recording and Analysis

ELMO PRICE

1. INTRODUCTION

The application of the fault recording function to the protective relay has considerably increased the availability of data for system disturbance analysis. It has broadened the range of fault-recording application to well beyond just the large critical substation at a reasonably low cost. Improvements in communications, the declining cost of technology, and the connectivity of intelligent electronic devices (IEDs) in the substation have made fault data readily accessible for quick and more intelligent analysis. Analysis programs utilizing digitally recorded fault data have proved to be valuable tools in analyzing protective relay and power system operations. New capabilities using distributed data allow new approaches to disturbance analysis of multiterminal data to provide more accurate solutions.

This chapter discusses digital fault-recording (DFR) capabilities of protective relays and basic approaches to analysis of their fault records. There are four parts. These are

1. A discussion of the relay digital fault record—important elements and how the event data is recorded.
2. Fault analysis tools
3. System fault analysis using real fault records illustrating the use of phasors and symmetrical components.
4. The industry standard COMTRADE format for data exchange.

2. THE DIGITAL FAULT RECORD

An adequate digital fault record (DFR) for protective relay operation and system analysis requires data recorded before, during, and after a system disturbance. There should be sufficient information in the form of analog and digital status (on–off) data to analyze and, if necessary, re-create the event. It may also be necessary to coordinate the relay operation with other devices around the system. The protective relay can do this. Its primary purpose, however, is to reliably clear the fault and restore the system with minimum outage time and system damage. Design limitations imposed by product cost limit the capabilities of functions secondary to protection, such as fault recording. These limitations, however, may be addressed by a better understanding of fault-recording usage and a more practical fault-recording design. The following discussion addresses data requirements for a protective relay fault record.

2.1 Analog and Digital (On–Off) Signals

2.1.1 Analog Signals

Analog signals are the digitally sampled voltage and current inputs into the relay. The sample frequency is limited by the relay design. The need to filter-out higher frequencies in the analog signals that are detrimental to accurate sampling measurement of the fundamental frequency is dependent on the relay's algorithm. Typically, a filtering

137

Table 10-1 Analysis Frequency and Resolution Based on Sample Frequency

Samples per cycle	Sampling frequency (Hz)	Nyquist limit frequency (Hz)	Digital analysis resolution (ms)
4	240	120	4.167
8	480	240	2.083
12	720	360	1.389
16	960	480	1.042
20	1200	600	0.833
32	1920	960	0.521
64	3840	1920	0.260

method is applied to limit the analysis frequency to half the sample frequency for antialiasing. Other considerations may also reduce the analysis frequency even more. Table 10-1 shows the Nyquist limit frequency for specified sample rates.

There are other techniques that allow the collection of sampled data before filtering is employed for the protection algorithm. In any case, the sample frequency directly affects the relay's memory requirements for recording and the frequency range that can be analyzed with the fault record.

Another factor to consider is the effect of sample frequency on the analysis being performed. Table 10-2 reflects the typical use of conventional (nonprotection) DFR data.

Eighty-five percent of the usage is for 60-Hz analysis. Of the remaining usage 8% was for harmonic analysis—determining the harmonic content of the analog signals. This analysis is generally limited to special applications, usually on the distribution system or major industrial load points. *Troubleshooting* is defined as checking the status of contacts, presence of voltages, sequence of

Table 10-2 Use of DFR Data

% of use	Type of analysis	Comments
70	Fault analysis	Fault type, location, sequence of events, and such
15	System parameter verification	Line impedance, fault current levels, and such
8	Harmonic analysis	Limited to distribution by respondents
7	Other	Switching transient analysis, Troubleshooting

events, and such, and is not generally frequency-dependent. It can be observed from the data that the Nyquist frequencies listed in Table 10-1 for analog data are adequate for most analytical uses, particularly for transmission line faults.

2.1.2 Digital (On–Off) Signals

Figure 10-1 is an example of a typical relay logic structure. The operating elements are the impedance, voltage, current, directional, or other measuring units that operate based on input analog signals and a set threshold value. They are either on or off. The binary inputs are voltage inputs into the relay to provide status information from external devices. The relay logic is usually modular in structure, utilizing inputs from the operating elements, binary inputs, or outputs from other logic modules. The individual relay outputs operate based on the assertion of specific logic signals.

The resolution of relay operating elements and logic status is generally provided at the interval the elements or logic is computed. If an element is computed once every cycle, then its resolution is approximately 16.67 ms. It is either on or off for the full cycle. If the element or logic is evaluated every sample, then the resolution is the period of the sample frequency. Different relay-operating elements can have different resolutions. An understanding of the resolutions of a particular relay's digital status data (operating elements), I/O, and logic, is important for analyzing the relay's performance. Table 10-1 shows the minimum digital signal analysis resolution (time between samples). It should be noted that the greater number of operating element and logic status signals available to analyze an operating process, the easier is the analysis.

2.1.3 Coordination of Digital and Analog Signals

The operating unit signal status is generally the result of a computation using several samples of data that are collected previous to the operating unit calculation. For example, it may be that an impedance unit is computed only once a cycle and that it uses sampled data collected in the previous cycle. This is illustrated in Figure 10-2. The shaded region shows the collected samples for the next impedance unit (Z) computation. The computation is made every cycle using the corresponding cycle of sampled data. This shows how the impedance unit output Z is asserted at some time after fault initiation and is deasserted at some time after the fault clears. This delay is dependent on the relay design and should be considered in the analytical process.

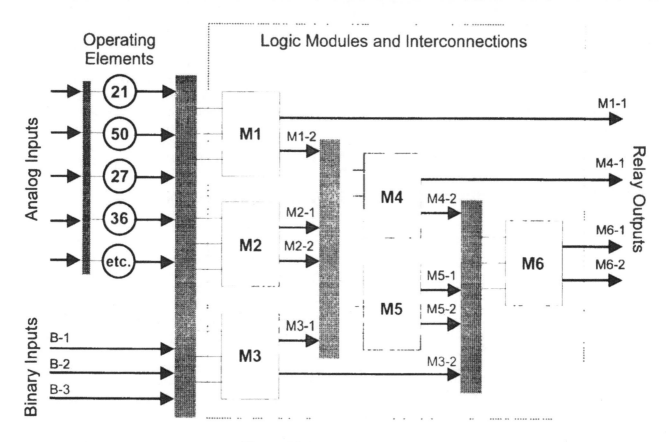

Figure 10-1 Relay logic structure.

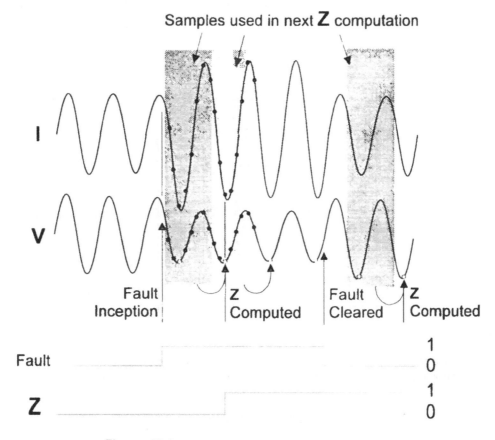

Figure 10-2 Coordination of digital and analog signals.

2.2 The Relay Fault Record

The following is provided to enable the protective relay engineer to develop an understanding of key fault record elements and how a fault record is produced in a relay. It is provided only as an example, as it will differ to some extent from equally good solutions provided by other relay manufacturers. There are four basic issues to contend with. These are the

1. System coverage
2. Event initiation detector
3. Pre- and post-initiation data requirements
4. Date and time of event

2.2.2 System Coverage

Because of limited relay memory it may be desirable to limit the fault records saved to faults on the protected line or in close proximity to the line. Collecting data for remote faults can quickly fill the memory buffers. There are several ways to restrict coverage utilizing relay-operating elements. A typical method might be to use a forward overreaching and a reverse impedance zone to define the boundary for saved fault records (Fig. 10-3). This ensures saving records for faults on the protected line, the next bus, and the reverse bus. The saving of a record may also be restricted to only those conditions for which the relay is

called to trip. In some cases, it may be desirable to just capture all remote faults possible, or those within the starting zone. The relay, however, should be checked often for data in the later case, as frequent recording may result in overwriting fault records. Most relays have settings to provide some degree of flexibility.

2.2.2 Event Initiation Detector

The event initiation detector, often called the starting unit or general start, is an overcurrent detector, undervoltage detector, rate-of-change detector, or other means to detect a fault inception or other event for which a fault record is desired. The rate-of-change detector has been used effectively to identify fault inception times and other sudden system parameter changes with one sample period resolution. The following is an explanation of the rate-of-change detector.

Figure 10-4 shows a typical periodic sine function being sampled at eight samples per cycle. In the general case, under steady-state conditions each sample's magnitude is measured as

$$S_K = A \sin\left(\frac{K2\pi}{N} + \theta\right)$$

where A is a constant, representing the peak amplitude of the function, θ is a constant depending on where sample 0

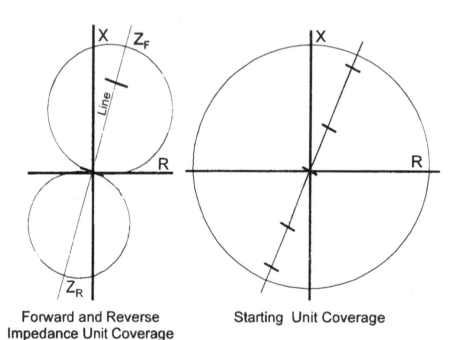

Forward and Reverse
Impedance Unit Coverage

Starting Unit Coverage

Figure 10-3 DFR coverage.

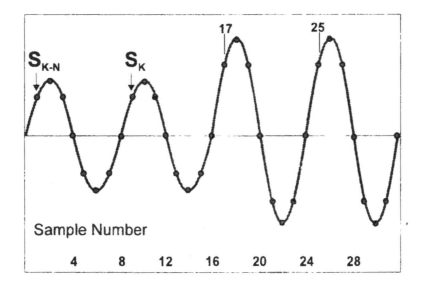

Figure 10-4   Operation of a rate-of-change detector.

is taken within the cycle, $K = 0, 1, 2, \ldots, N\text{-}1$, and N is the number of samples per cycle.

From one cycle to the next, the respective sample magnitudes S_K and $S_{K\text{-}N}$ are equal if there is no disturbance between them. The change detector continually compares the current sample S_K with $S_{K\text{-}N}$ one cycle back. If there is a difference greater than a specified value, then the change detector will operate. In this case, with a sample rate of 8 samples per cycle, the first change occurs when comparing sample 17 with 9. The change detector is asserted as the function transitions to the new steady state (fault) condition, and the difference between S_K and $S_{K\text{-}N}$ goes back to zero. This is indicated by the shaded area between samples 17 and 25. The steady-state fault condition will continue after sample 25 until there is another parameter change created by either a subsequent fault condition or fault clearing. Again the change detector will operate until a new steady-state operating condition is established. The resolution of the fault detection is the period ($1/f$) of the frequency at which the change detector is checked. This has proved to be a very reliable form of fault detection at a response less than 1.0 ms.

2.2.3   Pre- and Post-initiation Data Requirements

Preinitiation Data

Prefault (initiation) voltage and current data are the steady quantities that existed immediately before the fault inception. The prefault condition has not changed for a long time except for variations caused by load changes. These

changes are measured in seconds, rather than cycles and, therefore, are unseen. The cycle immediately preceding fault inception, as measured by a high-speed rate-of-change detector as described in the foregoing, will be the same as the prior 10 or 20 cycles. Therefore, minimum prefault data is required.

The relay's faulted phase selection and fault location algorithm, or the off-line fault data analysis program, defines the prefault data required. For correct analysis that requires the use of prefault phasor quantities computed from one cycle (period) of sampled data, it is necessary that the prefault samples have correct phase alignment with the fault cycle (period) of sampled data used in the calculation. The fault cycle used could be any cycle during the fault. Correct phase alignment requires that the two sample sets, prefault and fault, are taken from the same time points in their respective cycles. This is shown in Figure 10-5. If you were to start counting from 1 with the first prefault cycle sample up to and not including the first selected fault cycle sample and divide that number by the sample rate (samples per cycle) the remainder must be zero.

Two methods can be used to provide the required prefault data. The first is to use only one cycle of prefault data and rearrange the sample order by shifting the appropriate samples, as illustrated in Figure 10-6.

In this case of eight samples per cycle, the last five samples of the recorded prefault data enclosed in block A are equal to the last five samples in block B preceding the prefault record that were not recorded. The samples can be reordered before phasors are computed to produce the same result as correct phase alignment. The phasors can

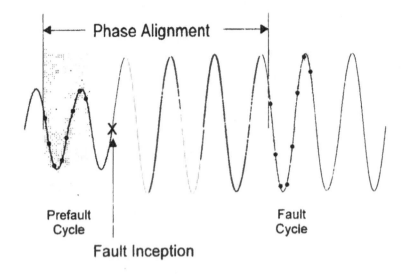

Figure 10-5   Prefault and fault cycle phase alignment.

also be directly computed from the samples of the prefault cycle, and then the appropriate phase shift can be added to provide the correct phase alignment.

The second method is to use two cycles of prefault data. With two cycles there will always be a full cycle of prefault data that will have correct phase alignment with any selected fault cycle.

Post-initiation Data

It is desired to capture the fault for its full duration from inception through clearing and reclosing. Forty-five cycles

is recommended for conventional DFR applications. For protective relays, this is not always practical to do with one continuous record. Correct zone-1 tripping for microprocessor relays would be expected to be in the two to three cycles. This is typically followed by two to four cycles of breaker clearing. Therefore, a seven-cycle post-initiation data record is adequate for most zone-1 or other three-cycle or fewer tripping function. One level greater would be to capture data for sequential tripping on the protected line. Sequential tripping is when a remote terminal clears a line-end fault and the local end responds to the system voltage and current changes and trips. An

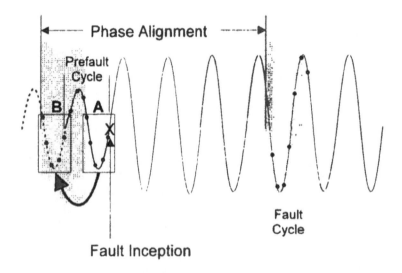

Figure 10-6   Altering the sample order for one-cycle prefault data.

example would be a loss-of-load trip function where the remote end trips and the local end responds after remote breaker clearing. If we consider that the trip logic and breaker trip will be sequential (one after the other) for the line terminal relays, 14 cycles of post-initiation should be considered. There is clearly no reason to save 45 or more cycles of data for zone-2 faults to capture line-end conditions or reclosing if there is a more practical way to obtain the needed information.

For a zone-2 fault, a record with pre- and post-initiation data can be recorded for the fault inception. The change detector will again operate on the breaker clearing at the end of zone-2 time and can be used to initiate a second record. Also, a reclose into a fault condition will produce an additional record. A successful reclose need not generate a fault record. There is no real need for the interim steady-state fault data except as a means for maintaining time coordination. The new generations of line relays are able to time-coordinate these records to less than a millisecond.

2.2.4 Date and Time

Date and time of the events can be time-tagged with an accuracy limited by the relay's sampling rate. The fault occurs at some time between two samples. The first sample after the fault operates the change detector. This sample initiates the record and is tagged with the time of the relay's clock. Multiple records from the same event can be coordinated using the time tags. The relay's clock is periodically set with an IRIG, GPS, or other synchronizing time signal to provide coordination with other system devices.

2.3 Recording the Fault Record

The following is a scenario for a typical fault recording using the four elements discussed in the foregoing. Once a fault inception has occurred a two-cycle prefault (initiation) record is held and post-fault data is collected for the next 14 cycles. The fault impedance is calculated at each post-fault sample during the 14 cycles and, if determined to be within the reach of the fault record zone, the 16 (2 pre- and 14 post-fault [initiation]) cycle record is time-tagged and saved. Change detection is not required while recording, but after the 14 cycles (the end of the record) the change detector is again monitored. If there are no changes (Fig. 10-7), the fault has cleared, the event is over, with a successful reclose.

If the faults still persist (Fig. 10-8), such as a zone-2 fault, the change detector will not operate until fault clearing (or other interim disturbance). At that time a second record for the zone-2 event is saved and time-tagged. In the event of reclosing, additional fault records will be recorded if the fault is still present.

3. FAULT ANALYSIS TOOLS

Much is to be learned by analyzing data for all faults, not just those faults for which operational problems occur. Protective relay fault-recording data, presented appropriately, can be used to

- Analyze the performance of the relay and the protection scheme
- Verify system (line) impedance parameters

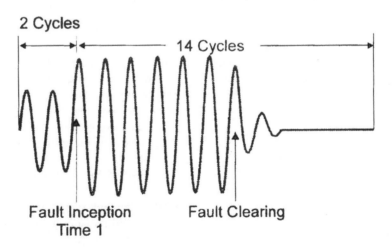

Figure 10-7 Typical single-record event.

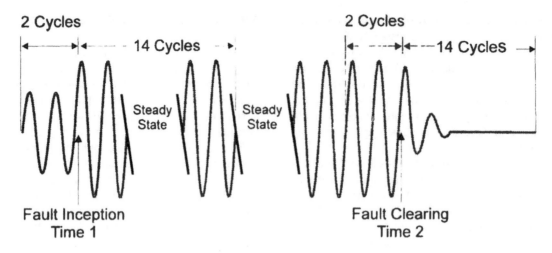

Figure 10-8 Typical two-record event.

- Provide more accurate fault location
- Aid in analyzing the power system response to system events
- Forecast potential problems
- Provide data for relay testing

It is with these uses in mind that analytical tool functions are defined and examples are provided for their use.

3.1 Graphical Display of Analog and Digital Data

A graphical display of analog and digital data as a function of time (samples) allows a quick visual review of the event. Fault and prefault analog magnitudes and relay element, key logic signals, and I/O operation can be easily observed. The analog plots should include all the phase voltage and current inputs, auxiliary inputs, such as transformer-polarizing current if used, and computed (from phase samples) zero- and negative-sequence quantities. An example is shown in Figure 10-9.

3.2 Multiple Record Analysis

The ability to analyze multiple records produced by all line-end terminal relays for the same event allows more accurate fault location, analysis of pilot operations, and analysis of more complex events.

3.3 Calculation of Fault Quantities

The phasors for all of the analog quantities should be computable for any one-cycle set of sampled data over the

length of the record. Correct phase aligned prefault phasors should also be computable for correct phase selection and fault location computations.

3.4 Calculation of Fault Impedance and Location

Fault impedance and fault location should be computable for any one-cycle set of sampled data over the length of the record. This should be done for each fault type: AG, BG, CG, ABG, BCG, CAG, AB, BC, CA, and ABC. Fault resistance should be calculated from multiterminal data records. The utilization of records from the line terminal relays recorded for the same fault allows accurate fault location and fault resistance calculations.

3.5 Phasor Plots

Plotting the voltage and current phasors, defined in the foregoing, on a polar plot to visually show their angular relationship is useful. This data can be useful when analyzing the performance of directional and other operating units, and records of two different relays that responded to the same event. An example is shown in Figure 10-10.

3.6 Plot of Fault Impedance Locus

Plotting the fault impedance locus over a specified range of the record provides the ability to determine the impedance location in the impedance (R,X) plane at a specified time and its effect on operating elements. This should be done for each fault type—AG, BG, CG, ABG, BCG,CAG, AB, BC, CA, and ABC. The intersection of the impedance

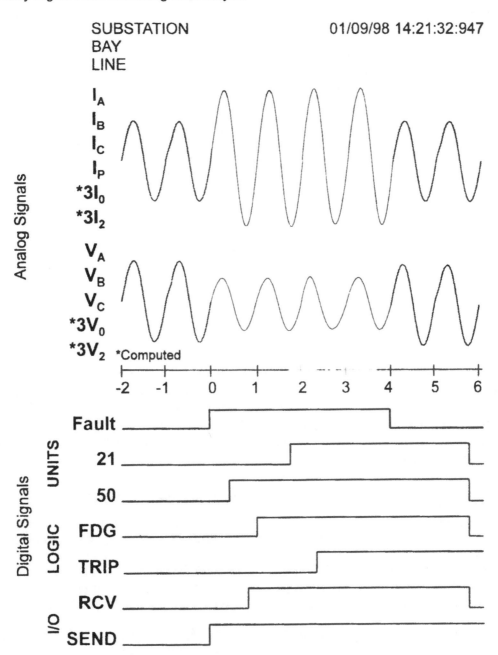

Figure 10-9   Typical fault-recording graphical display or printout.

locus and the set impedance operating characteristics can be observed for both faulted and nonfaulted phases. An example is shown in Figure 10-11.

3.7   Frequency Analysis

The ability to measure the frequency and harmonic components of the analog inputs should be provided, limited only by the protective relay filtering.

3.8   COMTRADE Files

It is important to provide standard COMTRADE output files for playback into computer-controlled test sets. With this capability, the fault recorded by the relay can be reproduced for additional testing analysis. Also other analysis software can use the COMTRADE file. Because of its importance a more detailed discussion of the COMTRADE file format follows.

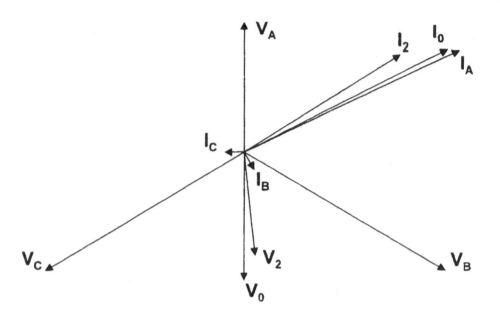

Figure 10-10 Typical phasor diagram for an AG fault.

4. COMTRADE

The COMTRADE Standard C37.111 defines a format for files containing transient waveform and event data collected from devices monitoring the power system or from computer models of power system simulators. The monitoring devices can be any of a large number of protective relays or digital fault recorders that save data in different file formats. The simulators are used to either test the relay directly or develop computer files with waveform data for relay testing with various available test sets. By defining a common format for these files, the COMTRADE Standard provides a link between many of these incompatible applications, enabling the interchange of fault, simulation, and test data.

The standard is for files stored on physical media, such as digital hard disk drives or diskettes. It is not a communications protocol for transferring data files over a communication network but rather, it defines a format that can be easily interpreted and implemented for data exchange.

The following discussion provides the reader an overview of the COMTRADE file format to show its structure and contents. The Standard C37.111 should always be referred to for development purposes.

4.1 Files Used in a COMTRADE Record

The 1998 revision of the COMTRADE Standard specifies a set of up to four files for storage and transport of tran-

sient recorded data. This file set makes up the COMTRADE record and includes a configuration, data, header, and information file. Only the configuration and data files are mandatory. The header and information files are optional.

4.1.1 Header File (*.HDR)

The header file is an optional ASCII text file of any length provided by the originator for the user to better understand the conditions of the transient record. The file has no specified format and, therefore, is not intended to be manipulated by an applications program. Examples of information, which may be included, are as follows:

- Description of the power system before disturbance
- Name of the station
- Identification of the line, transformer, reactor, capacitor, or circuit breaker that experienced the transient
- Length of the faulted line
- Positive- and zero-sequence resistance and reactance, capacitance
- Mutual coupling between parallel lines
- Locations and ratings of shunt reactors and series capacitors
- Nominal voltage ratings of transformer windings, especially the potential and current transformers
- Transformer power ratings and winding connections
- Parameters of the system behind the nodes where the data was recorded

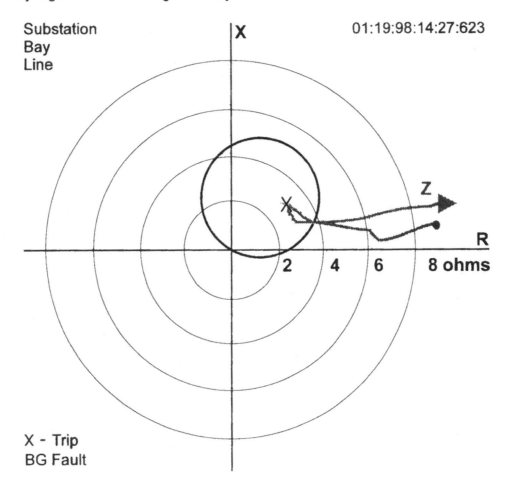

Figure 10-11 Typical impedance locus diagram.

- Description of how the data was obtained, whether it was obtained at a utility substation or by simulating a system condition on a computer program, such as EMTP
- Description of the antialiasing filters used
- Description of analog mimic circuitry
- The phase sequencing of the inputs
- Number of disks on which the record is stored

4.1.2 Configuration File (*.CFG)

The configuration file is a translation file for the data file; Table 10-3 summarizes such a file. It contains the information to restore the data file-sampled data to the most accurate representation possible of the original event. The configuration file has the following information on separate lines:

- Substation name, recording device ID, COMTRADE Standard revision year

Table 10-3 Example Configuration File

COMTRADE Sample,ID No 1,1998
10,6A,4D
01,Va,A,TEST,V,0.03,0,0.0,−10000,10000,1,1,S
02,Vb,B,TEST,V,0.03,0,0.0,−10000,10000,1,1,S
03,Vc,C,TEST,V,0.03,0,0.0,−10000,10000,1,1,S
04,Ia,A,TEST,I,0.03,0,0.0,−10000,10000,1,1,S
05,Ib,B,TEST,I,0.03,0,0.0,−10000,10000,1,1,S
06,Ic,C,TEST,I,0.03,0,0.0,−10000,10000,1,1,S
01,Zone-1,,,TEST,0
02,Zone-2,,,TEST,0
03,Zone-3,,,TEST,0
04,TRIP,,,TEST,0
60
1
2160,937
30/10/1998,00:00:00.000000
30/10/1998,00:00:00.266667
ASCII
1

- Total number of channels, number of analog channels with an "A" suffix, number of logic or status (1/0) channels with a "D" suffix
- Channel number, name, phase, circuit, units, scaling, offset, skew, minimum channel data value, maximum channel data value, transformer primary turns (ratio), transformer secondary turns (ratio), P or S for primary or secondary values of sampled data

  (A separate line is provided for each analog channel as specified with the foregoing A suffix. Number analog channels sequentially from 1.)
- Channel number, name, phase, circuit, quiescent state of input

  (A separate line is provided for each status channel as specified with the foregoing D suffix. Number status channels sequentially from 1.)
- System frequency
- Number of different sample rates in hertz used to record data in the record

  (This number is typically 1 for protective relays; 0 is used where the sample frequency is continuously variable.)
- Sample rate, the number of the last sample at the specified rate

  A separate line is provided for each sample rate or repeated sample rate of the record.
- Day/month/year (dd/mm/yyyy), hour:minutes:seconds (00:00:00:000000 resolution) time stamp for the first sample of the record
- Day/month/year (dd,mm,yyyy), hour:minutes:seconds (00:00:00.000000 resolution) time stamp for the sample that triggered or initiated the record
- Data type (ASCII or binary)
- Data file time stamp multiplication factor

  (Multiply this number times the time stamp integer of the data file to express the time stamp in microseconds. This number is normally 1, except for very long data files.)

It is important to clarify some terms used to define the data just defined.

Scaling and Offset

Scaling and offset are used to convert the recorded integer analog channel data to the appropriately scaled analog quantities. For protective relays these quantities are voltages and currents. These are sinusoidal functions the instantaneous values of which vary over a range of ± maximum peak value, equally offset from the zero value. The instantaneous values are processed by the A/D processor and converted to an integer within a range defined by the processor. The minimum and maximum integer values may

be equally offset from zero, or biased in one direction or another. This is the data that is recorded. Therefore the scaling and offset conversion factors are required to convert the integer analog data back to data that accurately represents the applied analog quantities. This is done using the equation

$$\text{Applied value} = \text{scaling} * (\text{integer data value}) + \text{offset}$$

Skew

It is desirable to sample all analog channels simultaneously to yield the most accurate representation of data. This is not always practical. The A/D processor may have a significant time differential between the first and last sample of all analog channels within the same sample period. The skew factor in microseconds is used to correct the sample time of a particular channel sampled at some time after the first channel sampled of that period. In general the skew error of protective relays is quite small, and a skew value of zero is generally used.

All data is not essential, but if not used, the space for it must be maintained with comma delimitation.

4.1.3 Data File (*.dat)

The data file retains all the analog and status channel data values and time for each sample of the transient data record. The values of the data are all integers and represent the original data after having been scaled by the factors defined in the configuration file; the following are specific data requirements for a data file record for each sample: Sample number; time in microseconds from the beginning of the record divided by the time stamp multiplication factor; analog channel 1 value, analog channel 2 value, . . ., analog channel N value; status channel 1 value, status channel 2 value, . . ., status channel N value ⟨CR/LF⟩.

The sample number of the first record should be 1 and all records listed sequentially by sample number. The numbers of analog and status channels are defined by the configuration file and are listed in the order therein defined. Status channel values are 1 or 0, representing the operated or not operated state, respectively. Each record is separated by the ⟨enter⟩ (CR/LF) key. An end of file marker is placed after the last record. This is basically no entry from the keyboard if developing the file from an ASCII text editor.

ASCII or Binary Data Files

ASCII data files, as shown in Table 10-4, can be easily read and imported to other software programs for analysis. Each data field of a sample record is separated by a comma and each sample record is separated by ⟨CR/LF⟩. This presents an easy to understand text representation of the data

Table 10-4   Example ASCII Data File, Based on Table 10-7

0001,0000000000,000573,−003101,002527,000016,−000089,000072,0,0,0,0	Start
0002,0000000463,001128,−003250,002121,000032,−000093,000060,0,0,0,0	
0003,0000000926,001649,−003300,001649,000047,−000095,000047,0,0,0,0	
0004,0000001389,002121,−003250,001128,000060,−000093,000032,0,0,0,	
0662,0000306019,001212,001128,−003250,000188,000032,−000093,0,0,0,0	
0663,0000306481,000942,0016490,−003300,000199,000047,−000095,0,0,1,0	Zone 3
0664,0000306944,000644,002121,−003250,000204,000060,−000093,0,0,1,0	
0665,0000307407,000327,002527,−003101,000204,000072,−000089,0,0,1,0	
0666,0000307870,−000001,002857,−002858,000197,000081,−000082,0,0,1,0	
0720,0000332870,000000,−002858,002857,−000179,−000082,000081,0,0,1,0	
0721,0000333333,000327,−003101,002527,−000168,−000089,000072,0,0,1,0	
0722,0000333796,000644,−003250,002121,−000152,−000093,000060,0,0,1,1	Trip
0723,0000334259,000942,−003300,001649,−000131,−000095,000047,0,0,1,1	
0724,0000334722,001212,≅003250,001128,−000105,−000093,000032,0,0,1,1	
0935,0000432407,−000246,−001084,003100,−000468,000267,000088,0,0,0,0	
0936,0000432870,000000,−001225,002857,−000454,000330,000081,0,0.0.0	
0937,0000433333,000245,−001329,002527,−000426,000382,000072,0,0,0,0	

that can be easily imported to many spreadsheet applications with minimum conversion effort. The ASCII files may be large, requiring excessive amounts of disk space for storage and long transmission times when transferring data from one location to another.

Binary data files are basically unreadable in a test format and require a computer program to translate the data into an interpretable format. They do offer considerable economy of disk storage requirements and communication times compared with ASCII files. This is preferable if the files are quite large. The binary data file is a continuous stream of fixed-length data fields. Each field is defined by its sequential position within the file. The sequence is sample number (4 bytes), sample time (4 bytes), channel data (2 bytes each channel in sequence, as specified in the configuration file), next sample number, next sample time, next set of channel data, and so on.

4.1.4   Information File (*.INF)

The optional information file provides for the exchange of information about the event recorded in the COMTRADE record, which may enable enhanced manipulation or analysis of the data. This optional information is stored in a separate file to allow full backward and forward compatibility between current and future programs that utilize COMTRADE files.

The file format is similar to the *.INI file used in the Windows operating environment. These files provide specific information of the application for operation in the Windows environment.

Some of the sections in the information file may duplicate information stored in the configuration file (*.CFG).

This is permissible, but under no circumstances is the duplicate data to be omitted from the configuration file.

4.2   Application of the COMTRADE Format

The following gives several examples of using event data, which is facilitated by availability of the COMTRADE standard file format.

4.2.1   Event Analysis with Universally Accessible Data

Digital fault recorders, protective relays, and other intelligent electronic devices that monitor devices and systems with sampled analog and status (on–off) data will save event data in a specific file format within the device's memory. The file format will vary depending on device memory, file compression requirements, event time, sampled frequency, and such. This data is either stored in the device with the COMTRADE file format or transferred to another location, such as a remote computer, where the file is converted to the COMTRADE file format. The data is now available for application analysis tools that can read the COMTRADE file format.

4.2.2   Protective Relay Testing by Re-creating Conditions from Recorded Data

In this application, fault data from intelligent electronic devices, such as fault recorders, relays, or from transient simulation programs, such as EMTP, are downloaded to the waveform memory of a microprocessor-controlled relay test system and the records are reproduced at secondary injection voltage and current levels.

4.2.3 Automated Data Collection and Management

Common file formats enable automated collection of data from different remote fault-recording devices at a master location for automatic event coordination and analysis. All records for the same event are identified, synchronized, analyzed, and saved. This requires automated data file recognition and translation for different product manufacturers. Data management will also be required to organize the data and provide event summaries.

Not only are individual events analyzed, but also a historical relational database can be maintained. This permits statistical analysis of fault location, fault type, protection scheme performance, and other parameters possibly revealing areas to improve system operation.

4.2.4 Protection-Testing Database

A file of fault records of actual system events that produce both desirable and undesirable relay operations, or test the performance boundaries of the relay, can be developed, grown, and maintained. These records can be used with modern test equipment using the COMTRADE file format to re-create the event during product development by the manufacturer or product evaluation by the user to reduce product acceptance time.

4.2.5 Functional System Testing

Predefined test cases may be developed for functional testing of specific system protection applications. For example, end-to-end testing of a transmission line pilot protection scheme can be done with the use of synchronous secondary injection tests at each terminal (substation) facilitated by GPS public domain timing signals. Test cases can be prepared using simulation tools such as EMTP to verify correct installation, pilot channel coordination, reclosing, settings, and other details pertinent to the application.

5. FAULT ANALYSIS EXAMPLES

Four examples are discussed to show different examples of fault analysis. The examples are taken from fault data recorded by the first-generation line protection product. Any problem discussed that resulted in an undesired operation has been corrected. Also, rather than providing detailed analysis of each example, only the key points of the analysis are discussed to illustrate the power of fault analysis tools.

5.1 Example 1. Potential Circuit Grounding

This problem involved two numerical relays providing redundant protection at the same line terminal, as shown in Figure 10-12. The backup relay consistently computed a fault location for single phase-to-ground faults farther than the actual fault location, whereas the primary relay was consistently accurate. The analog data from the two relays for one specific event were plotted and superimposed, as shown in Figure 10-13. Immediate inspection revealed a zero-sequence voltage problem. The current and voltage phasors were then computed from the same cycle of fault data. The results are shown in Table 10-5.

As can be observed from Table 10-5, all current quantities compare excellently. This rules out a current circuit problem. Likewise the positive ($V1$) and negative ($V2$) sequence voltages compare excellently. This rules out a

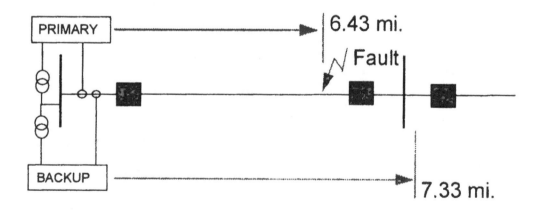

Figure 10-12 Redundant protection with different fault location computations.

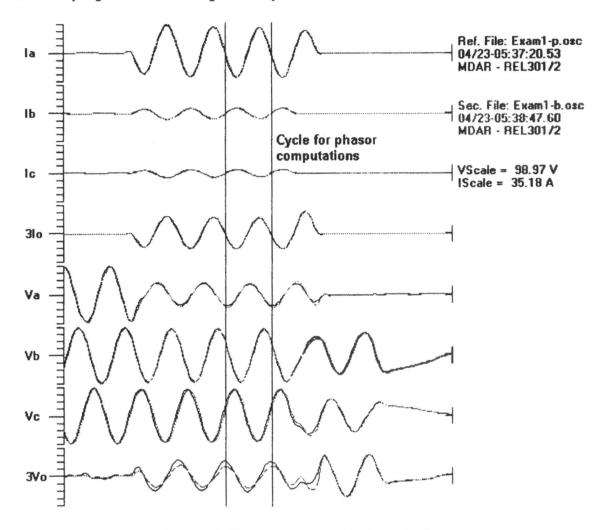

Figure 10-13 Plot of primary and backup analog data.

phase voltage connection problem. This leaves the neutral of the backup potential circuit as suspect. The difference in the zero sequence voltage (ΔV_0) of the primary and backup relays was computed by:

$$\Delta V_0 = \frac{[3V_0(\text{primary}) - 3V_0(\text{backup})]}{3}$$

$$\Delta V_0 = \frac{(35.71 \angle 197° - 24.64 \angle 192°)}{3} = 3.79 \angle 208°$$

Addition of ΔV_0 to each backup phase voltage clearly gives comparable quantities seen and recorded by the primary relay (Table 10-6). Therefore, something is generating this voltage in the backup potential circuit.

The two relays were provided voltage from separate secondary windings of one potential transformer. Inspec-

Table 10-5 Comparison of Phasors

Phasor quantity	Primary	Backup
Va	27.76 $\angle$ 0°	31.01 $\angle$ 3°
Vb	67.64 $\angle$ 251°	64.9 $\angle$ 253°
Vc	66.45 $\angle$ 126°	66.31 $\angle$ 123°
$3V_0$	35.71 $\angle$ 197°	24.64 $\angle$ 192°
V_1	53.82 $\angle$ 7°	53.87 $\angle$ 7°
V_2	14.66 $\angle$ 193°	14.86 $\angle$ 194°
I_a	22.35 $\angle$ 284°	22.23 L 284°
I_b	4.82 $\angle$ 101°	4.82 $\angle$ 101°
I_c	3.45 $\angle$ 100°	3.45 $\angle$ 100°
$3I_0$	14.08 $\angle$ 286°	13.99 $\angle$ 287°
I_1	8.84 $\angle$ 281°	8.8 $\angle$ 281°
I_2	8.83 $\angle$ 286°	8.78 $\angle$ 287°

Table 10-6 Effect of ΔV_0

Phasor	Backup	ΔV_0	Backup + ΔV_0	Primary
V_a	31.01 ∠ 3°	3.79 ∠ 208°	27.61 ∠ 0°	27.76 ∠ 0°
V_b	64.9 ∠ 253°	3.79 ∠ 208°	67.63 ∠ 251°	67.64 ∠ 251°
V_c	66.31 ∠ 123°	3.79 ∠ 208°	66.72 ∠ 126°	66.45 ∠ 126°

tion of the potential circuit revealed multiple grounds on each circuit, one at the voltage transformer and one at the relay panel. This condition will most certainly produce a potential difference between the ground points owing to ground potential rise during ground faults. This problem is addressed in the *IEEE Guide for the Grounding of Instrument Transformer Secondary Circuits* in which single-point grounding is recommended at the first point of application of secondary potential circuit in the control house (panel).

5.2 Example 2. Incorrect Directional Unit Operation on a Resistance-Grounded System

Figure 10-14a shows the operating region of the forward and reverse directional units, as defined by the zero-sequence voltage and current phasor relationship during a ground fault. A ground fault is defined as forward when I_0 leads V_0 by 30°–210°. The normal region of operation de-

fines the region where I_0 is expected, relative to V_0, for ground faults on effectively grounded transmission systems at all voltage levels. Figure 10-14b shows I_{0R} and V_0 relationship for a reverse ground fault that occurred on a resistance grounded cable transmission system. This resulted in the forward directional unit operation and an undesired trip. A review of many fault records of forward (I_{0F}) and reverse (I_{0R}) faults at several locations showed that I_0 occurred close to the balance point (zero-torque line) between forward and reverse.

Analysis showed that the transformer neutral-ground resistors not only introduced resistance into the zero-sequence network, but also caused the cable shunt capacitance to become significant. This caused a phase shift in the resulting I_0 from the normal range to that observed in Figure 10-14b.

The solution was simply to rotate the zero-torque line, the boundary between forward and reverse operation, as shown in Figure 10-14b. This enabled correct directional sensing for this unique system.

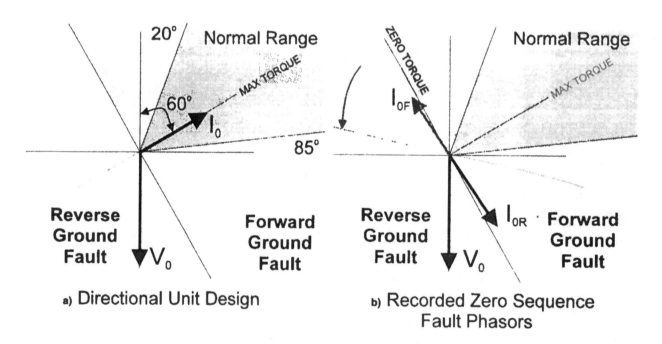

Figure 10-14 Directional unit operation.

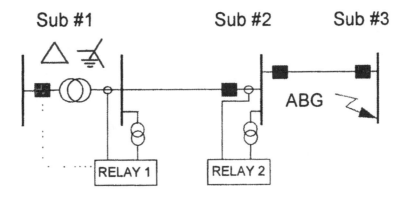

Figure 10-15   Isolated ground source at Sub 1.

5.3   Example 3. Misoperation Caused by Isolated Ground Source

The following example discusses a misoperation that was the result of undesired, but real, operating condition. The system as shown in Figure 10-15 was set up when the breaker, as indicated at Sub 1, was opened. This left only a zero-sequence source for the Sub 1 to Sub 2 line at Sub 1.

When the two phase-to-ground faults occurred at Sub 3, approximately 0.4 secondary amperes of zero-sequence current flowed in the line. The fault type selections of both Relay 1 and Relay 2 were three-phase. The phase selector could not determine any fault type and defaulted to a three-phase fault because of the small, and relatively equal, phase current magnitudes. The operation of the zone-1 phase unit requires only one of three three-phase units to operate with three-phase fault type selection. By using the following three-phase unit equations and data of Table 10-7, analysis shows a three-phase unit operation to occur at both relays.

$V_0 = V_{XG} - I_X Z_R$ operating

X = A, B, C; and ZR = relay impedance setting

$V_Q = V_{XY}$ polarizing

Table 10-7   Fault Voltages

Phase voltage	Relay 1		Relay 2	
	mag.	Angle	Mag.	angle
A	10.5	0	9.7	0
B	10.9	311	10.3	311
C	62.5	204	62.5	208
BC	60.2	214	61.0	217

XY = CB for A, AC for B, and BA for C phase units

The unit operates where V_0 leads V_Q. Neglecting the effect of current, it is quickly seen that the operating quantity V_A leads the polarizing quantity V_{BC} for both relays. These satisfy that the relays will trip. Relay 2 DFR data, shown in Figure 10-16, shows the zero-sequence current, the three-phase fault type selection multiphase (MP) signal, and the zone-1 three-phase logic pickup ZIP. A similar record for relay 1 shows the same results.

The most immediate solution to the problem was to enable supervision of the three-phase–tripping logic with phase overcurrent units set to an appropriate value. Three-phase trip logic can also be modified requiring all three-phase units to operate, all three-phase directional units to operate, or trip blocking when zero-sequence current is present.

Although this is not a complex problem, it illustrates how a protective relay fault record allows a quick visual analysis of analog and digital quantities to point quickly to what elements need to be studied.

5.4   Example 4. Dual Polarization and Unbalanced Phase Voltages

Dual polarization is the utilization of two directional sensing units. The first directional unit uses a polarization current source external to the relay and the internally derived zero-sequence current. The external polarization source provides greater sensitivity to ground faults and is usually a ground current source, such as a wye-grounded–delta power transformer or other grounding transformer. The second directional unit uses internally derived zero-sequence current and voltage. Both units' directional characteristics are defined in Figure 10-17.

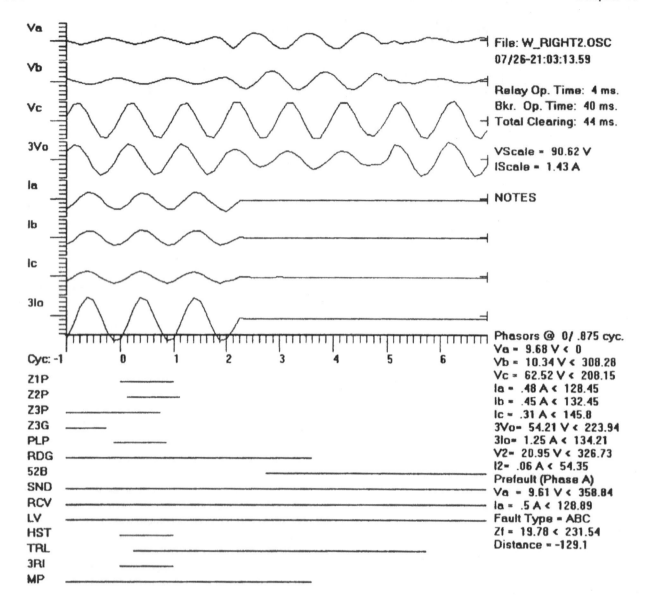

Figure 10-16 Relay 2 digital fault record.

The dual polarization directional logic was such that either unit could enable directional tripping or pilot logic. The externally polarized I_P unit would provide the sensitivity and the internal $3V_0$ unit would serve as backup if the I_P source was removed from service.

Incorrect pilot tripping occurred on a typical two-terminal transmission line protection scheme for a distant remote fault, as illustrated on Figure 10-18. The data from both terminal relays indicated a forward fault that enabled pilot tripping. A review of the appropriate phasor relationships and the operating characteristics defined in Figure 10-17 showed forward $I_P/3I_0$ and reverse $3V_0/3I_0$ operation at one terminal, and reverse $I_P/3I_0$ and forward $3V_0/3I_0$ operation at the other terminal. The fault quantities from one terminal are shown in Table 10-8. I_0 leads V_0 by 203° (112 + 91), therefore the $3V_0/3I_0$ unit is forward. I_0 leads I_P by 191° (112 + 79), therefore the $I_P/3I_0$ unit is reverse. Because one unit sees the fault as forward at both terminals, pilot tripping resulted. Both unit types, individually as a two-terminal system, operated correctly at each terminal—one forward and one reverse, for the external fault. Operating together, however, resulted in erroneous operation.

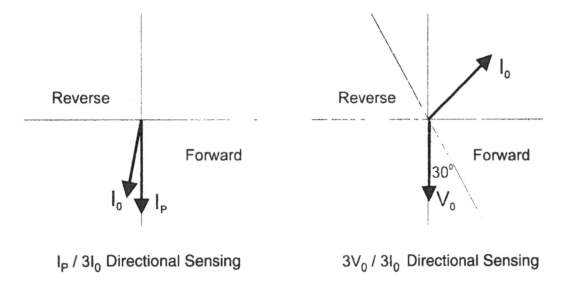

$I_P / 3I_0$ Directional Sensing $3V_0 / 3I_0$ Directional Sensing

Figure 10-17 Dual polarizing directional characteristics.

Therefore, analysis determining why the two units differ in directional sensing for this fault was needed.

Table 10-8 also shows an unbalance in the nonfaulted phases. This unbalance, although small, introduces enough effect on the "expected" zero-sequence voltage angle at these small values to cause the forward directional sensing. This is illustrated by changing the V_{ANG} values of

Table 10-6 for phases B and C from $-116°$ and $119°$ to $-120°$ and $120°$, respectively, and computing the zero-sequence voltage. This results in zero-sequence voltage angle of 184° which is very close to what is normally expected. For this illustration, I_0 lags V_0 by 72° $(112 - 184)$, which results in a reverse operation by the $3V_0/3I_0$ directional unit and in agreement with the $I_P/3I_0$ unit.

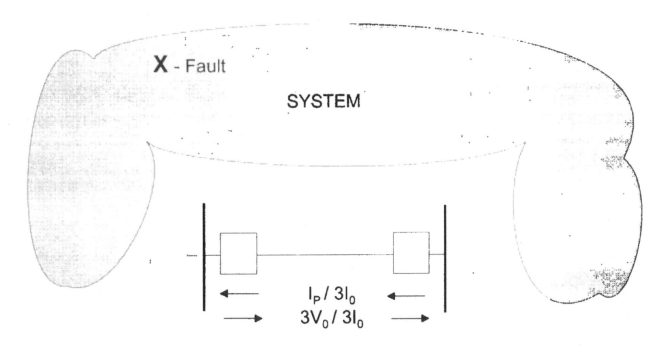

Figure 10-18 External single phase-to-ground fault.

Table 10-8 Recorded Fault Quantities and Directional Unit Operation

Quantity	V_{MAG}	V_{ANG}	I_{MAG}	I_{ANG}	$I_{P\text{-}MAG}$	$I_{P\text{-}ANG}$
Phase A	59.6	0°	5.0	133°	—	—
Phase B	64.8	−116°	2.0	72°	—	—
Phase C	64.4	119°	2.3	−55°	—	—
Zero-Sequence	0.64	−91°	1.32	112°	5.5	−79°

Directional Unit	$I_{0\text{-}ANG}$ leads reference	Operation
Zero-Sequence $3V_0/3I_0$	$I_{0\text{-}ANG}$ −$V_{0\text{-}ANG}$ 203°	Forward
Ext. current pol. $I_P/3I_0$	$I_{0\text{-}ANG}$ −$I_{P\text{-}ANG}$ 191°	Reverse

It should be clearly understood that the $3V_0/3I_0$ units at both terminals are equally affected by the unbalance and operate correctly as a system—one forward and one reverse. Likewise the $I_P/3I_0$ units operate correctly, but are immune to any unbalanced voltages. Applied together in dual-polarizing logic, however, the discussed operating differences need to be considered.

The solution is to provide logic that detects the presence of the external polarizing current and giving the $I_P/3I_0$ unit operating priority. The $3V_0/3I_0$ directional unit can operate only if there is no external polarizing current. This implies that the polarizing current source, usually a transformer, has been removed from service. This contingency still leaves the exposure to the same problem until the settings are changed. This, however, is better than having no polarization source at all.

Bibliography

LIANCHENG WANG

GENERAL

Blackburn JL, ed. Applied Protective Relaying. Westinghouse Electric Corporation, 1982.

Blackburn JL. Protective Relaying. New York: Marcel Dekker, 1987.

Elmore WA, ed. Protective Relaying Theory and Application. New York: Marcel Dekker, 1994.

IEEE Committee Report. IEEE Bibliography of Relay Literature. PE-416-PWRD-0-08-1998.

Wagner CF, RD Evans. Symmetrical Components. New York: McGraw-Hill, 1961.

COMMUNICATION FUNDAMENTALS

Driscoll FF. Data Communications. New York: Saunders College Publishing, 1992.

Haus RJ. Fiber Optics Communications Design Handbook. Englewood Cliffs, NJ: Prentice Hall, 1980.

IEEE Tutorial Course. Advancements in Microprocessor Based Protection and Communication. IEEE Catalog No. 97TP120-0, Chap. 9, Relay communications.

Ray RE. Fiber Optic Communications for Utility Systems. Coral Springs, FL: Pulsar Technologies, Technical Publication 95-1.

CURRENT DIFFERENTIAL RELAYING

AIEE Committee Report. Protection of pilot-wire relay circuits. AIEE Trans 78:205–212, 1959.

Calero JF, WA Elmore. Current differential and phase comparison relaying schemes. 46th Annual Georgia Tech Protective Relaying Conference, Atlanta, Georgia, 1992.

Harder EL, MA Bostwick. Pilot wire protection circuits. AIEE Trans 61:645–652, 1942.

Hinman WL. Pilot protection of transmission lines: distance based vs current-only. 50th Annual Georgia Tech Protective Relaying Conference, Atlanta, Georgia, May 1–3, 1996.

Patterson H. High speed phase segregated line differential relays. Proceedings of CEA Engineering and Operating Div. Meetings, March 26–31, 1995.

Sun SC, RE Ray. A current differential relay system using fiber optics communications. IEEE Trans Power Apparatus Sys PAS-102:410–419, 1983.

PILOT CHANNELS FOR PROTECTIVE RELAYING

AIEE Committee Report. Guide to the application and treatment of channels for power line carrier. AIEE Trans 73(pt III):417–436, 1954.

Dzieduszko J. On the upcoming revolution in protection systems. Texas A&M Conference, April, 1996.

IEEE Committee Report. Power line carrier practices and experiences. IEEE Trans Power Deliv 10:639–646, 1995.

IEEE. IEEE Guide for Power-Line Carrier Applications. IEEE Standard 643–1980.

IEEE Committee Report. Survey of optical channels for protective relaying practices and experiences. IEEE Trans Power Deliv 10:647–658, 1995.

IEEE Protective Relay Applications of Audio Tones Over Telephone Channels. IEEE Standard No. 305, 1969.

IEEE Power System Relaying Committee Report. Fiber optic channels for protective relaying. IEEE Trans Power Deliv 4:165–176, 1989.

Lensner HW. Protective relaying over microwave channels. AIEE Trans, 71(pt III):240–245, 1952.

Perz MC. Natural modes of power line carrier on horizontal three-phase lines. IEEE Trans PAS July: 679–691, 1964.

Ray RE. Channel considerations for power-line carrier. ABB RPL 83-3.

Ray RE. Fiber-optic communications for utility systems. Georgia Tech Relay Conference, April 1993.

Sanders M, RE Ray. Power line carrier channel and application considerations for use with transmission line relaying. 50th Annual Georgia Tech Protective Relaying Conference, Atlanta, Georgia, 1996.

TRANSMISSION LINE PILOT RELAYING

AIEE Committee Report. Protection of pilot wire relay circuits. AIEE Trans 78:205–212, 1959.

Dzieduszko JW. Combined-sequence phase comparison relaying. 51th Annual Georgia Tech Protective Relaying Conference, Atlanta, Georgia, April 30–May 2, 1997.

Elmore WA. Current differential and phase comparison relaying compared with pilot distance schemes. 25th Annual Western Protective Relay Conference, Spokane, Washington, October 13–15, 1998.

IEEE C37.113, Power System Relaying Committee Guide for protective relay applications to transmission lines. 1998.

Killen RB, GG Law. Protection of pilot wires from induced potentials. AIEE Trans 65:267–270, 1946.

Quest R, JW Dziedusko, R Hedding. Field experience with segregated phase comparison protection system. 49th Annual Conference for Protective Relay Engineers, College Station, Texas, April 15–17, 1996.

THREE-TERMINAL LINE PROTECTION

Alexander GE, JG Andrichak. Application of distance relays to three terminal lines. 22th Annual Western Protective Relay Conference, Spokane, Washington, October 24–26, 1995.

IEEE. Guide for Protective Relaying of Utility–Consumer Interconnections. IEEE C37.95-1989. (ANSI).

IEEE Protection Aspects of Multi-Terminal Lines. IEEE-PES Special Publication. 79TH0056-2-PWR, IEEE Service Center.

Kobayashi J, Y Ohura, M Yuki, K Hashisako, K Seo, K Suzuki, A Tsuboi, Andow F. The state of the art of multi-circuit and multi-terminal overhead transmission line protection systems associated with telecommunication systems. CIGRE General Session, paper No. 34-201, 1990.

Rook MJ, LE Goff, GJ Potochnry, LP Powell. Application of protective relays on a large industrial-utility tie with industrial co-generation. IEEE Trans PAS PAS-100:2804–2812, 1981.

PROGRAM DESIGN FOR MICROPROCESSOR RELAYS

Calero F. Development of a numerical comparator for protective relaying: Part I. IEEE Trans Power Deliv, 11:1266–1273, 1996.

Elmore WA, F Calero, Y Yang. Evolution of distance relaying principles. 49st Annual Georgia Tech Protective Relay Conference, Atlanta, GA, May 3–5, 1995.

Kennedy JM, GE Alexander, JS Thorp. Variable digital filter response time in a digital distance relay. 48th Annual Georgia Tech Protective Relaying Conference, Atlanta, GA, 1994.

Phadke AG, JS Thorp. Computer Relaying for Power Systems. New York: John Wiley & Sons, 1988.

Proakis JG, DG Manolakis. Digital Signal Processing: Principles, Algorithms, and Applications. 3rd ed. Englewood Cliffs, NJ: Prentice Hall, 1996.

Sonnemann WK, HW Lensner. Compensator distance relaying: I. General principles of operation. AIEE Trans 77(pt III):372–382, 1958.

Udren EA, HJ Li. Transmission line relaying with microprocessors. 41st Annual Conference for Protective Relay Engineers, College Station, TX, April 12–14, 1993.

Wang L. Frequency response of phasor-based microprocessor relaying algorithms. IEEE Trans Power Deliv 14:98–105, 1999.

Wang L, E Price. High-speed microprocessor distance relaying for transmission lines. 25th Annual Western Protective Relay Conference, Spokane, WA, October 13–15, 1998.

SERIES-COMPENSATED LINE PROTECTION PHILOSOPHIES

Anderson F, WA Elmore. Overview of series compensated line protection philosophies. 17th Annual Western Protective Relay Conference, Spokane, WA, October, 1990.

Andrichak JG, GE Alexander, WZ Tyska. Series compensated line protection: practical solutions. 17th Annual Western Protective Relay Conference, Spokane, WA, October 22–25, 1990.

Coney RG, GH Topham, MG Fawkes. Experience and problems with the protection of series compensated lines. 4th International Conference on Developments in Power System Protection. IEE publication No. 302, 1989, pp 177–181.

Cheetham WJ, A Newbould, G Stranne. Series compensated line protection: system modelling and test results. 15th Annual Western Protective Relay Conference, Spokane, WA, October 25–27, 1988.

Goldsworthy DL. A linearized model for Mov-protected series capacitors. IEEE Trans Power Sys 2:953–958, 1987.

Jaysasinghe J, RK Aggarwal, AT Johns, ZQ Bo. A novel non-unit protection for series-compensated EHV transmission lines based on fault generated high frequency voltage signals. IEEE Trans Power Deliv 13:405–413, 1998.

Marttila R. Protection of series capacitor equipped lines. IEEE Trans PAS 7: July 1992.

SINGLE-POLE TRIPPING

Elmore WA. Some thoughts on single-pole tripping. 49th Annual Conference for Protective Relay Engineers. College Station, TX, April 15–17, 1996.

Fitton DS, RW Dunn, RK Aggarwal, AT Johns, A Bennett. Design and implementation of an adaptive single pole autoclosure technique for transmission lines using artificial neural networks. IEEE Trans Power Deliv 11:746–756, 1996.

IEEE Committee Report. Single phase tripping and auto reclosing of transmission lines. IEEE Trans Power Deliv 7:182–192, 1992.

IEEE Transmission and Distribution Committee Report. Bibliography on single pole switching. IEEE Trans PAS 94:1072–1078, 1975.

Jackson B, MR Best, RH Bergen. Application of a single pole protective scheme to a double-circuit 230 kV transmission line. 25th Annual Western Protective Relay Conference, Spokane, Washington, October 13–15, 1998.

SUBSTATION AUTOMATION AND RELAY COMMUNICATIONS

Ackerman WJ. Current Trends in Substation Automation. ABB Power T&D Company Inc., 1996.

Adamiak MG, RC Patterson. Communication requirements for protection and control in the 1990's. 44th Annual Texas A&M Protective Relaying Conference, College Station, Texas, April 15–17, 1991.

Garrity JP. Data collection and control techniques for protective relays. 43rd Annual Texas A&M Protective Relaying Conference, College Station, Texas, April 23–25, 1990.

IEEE Tutorial. Communication Protocols. Catalog Number 95-TP-103, IEEE Service Center, 445 Hoes Lane, Piscataway, NJ 08854.

IEEE Std C37.1-1994. Definition, Specification, and Analysis of Systems used for Supervisory Control, Data Acquisition, and Automatic Control. IEEE Service Center, 445 Hoes Lane, Piscataway, NJ 08854.

Stallings W. Networking Standards, A Guide to OSI, ISDN, LAN, and MAN Standards. New York: Addison Wesley, 1993.

Substation Integrated Protection, Control and Data Acquisition Phase 1, Task 2 Requirement Specification. RP3599-01, Preliminary Report, Version 0.4, March 1, 1996.

PROTECTIVE RELAY DIGITAL FAULT RECORDING AND ANALYSIS

ABB Power T&D Company Inc., Substation Control and Communications Survey. 1992.

ANSI/IEEE. Standard Common Format for Transient Data Exchange (COMTRADE). C37.111, 1991.

ANSI/IEEE IEEE Guide for the Grounding of Instrument Transformer Secondary Circuits and Cases. C57.13.3, 1983.

IEEE Power System Relaying Committee. Fault and disturbance data requirements for automated computer analysis. Special Publication, IEEE Catalog No. 95 TP 107. IEEE.

IEEE standard common format for transient data exchange. IEEE C37.111, 1999.

Henville CF. Digital relay reports verify power system models. IEEE Trans Power Deliv 13:386–393, 1998.

Price E. Protective relay digital fault recording and analysis. 25th Annual Western Protective Relay Conference, Spokane, Washington, October 13–15, 1998.

Index

T - #0343 - 101024 - C0 - 280/208/10 [12] - CB - 9780824781958 - Gloss Lamination